Acquisition and Analysis of Terrestrial Gravity Data

Gravity surveys, detecting minute variations of the gravitational field at the Earth's surface, have a huge range of applications, indicating density variations in the subsurface, identifying man-made structures, characterizing local changes of rock type, or even mapping deep-seated structures at the crust/mantle boundary. This important one-stop book combines an introductory manual of practical procedures with a full explanation of analysis techniques, enabling students, geophysicists, geologists, and engineers to fully understand the methodology, applications, and limitations of a gravity survey.

Filled with examples from a wide variety of acquisition problems, the book instructs students in avoiding common mistakes and misconceptions. The authors also explore the increasing near-surface geophysical applications being opened up by improvements in instrumentation, and provide some more advanced-level material to give a useful introduction to potential theory.

Springing from the authors' extensive teaching experience and professional participation in a wide range of engineering applications, this book shares their considerable insight into data-acquisition procedures and analysis techniques. A key text for graduate students of geophysics, this is also an invaluable resource for professionals using gravity surveys, from civil engineers and archaeologists, to oil and mineral prospectors and geophysicists seeking to learn more about the Earth's deep interior.

Leland Timothy Long is Emeritus Professor of Geophysics at the Georgia Institute of Technology, where he has taught geophysics for more than 40 years, on topics including exploration geophysics, potential theory, seismology, data analysis, and earth science. He is also a licensed Professional Geologist in Georgia and consults on topics of gravity data interpretation, seismic hazard, and blast vibration. Professor Long's research interests in potential methods include the interpretation of regional crustal structures for seismic hazards, deflections of the vertical, statistical properties of gravity fields, and microgravity for the detection of sinkholes. His research interests in seismology have emphasized seismicity, seismic networks, induced seismology, many aspects of theoretical seismology related to scattering and scattering inversion, and near-surface surface wave analysis. In 2006, Professor Long was awarded the Jesuit Seismological Association Award honouring outstanding contributions to observational seismology.

Ronald (Ron) Kaufmann is President of Spotlight Geophysical Services in Doral, Florida and has more than 19 years of geophysical consulting experience, including positions of Vice President and Senior Geophysicist at Technos, Inc. He is a licensed Professional Geophysicist in California and a Professional Geologist in Tennessee. Mr Kaufmann is an expert in the use of microgravity for karst and geologic hazard investigations and has personally acquired gravity measurements at more than 20,000 stations for a variety of near-surface investigations. He has led geophysical projects for environmental, geotechnical, and groundwater applications across the US and Latin America, including several comprehensive geophysical investigations for the Panama Canal Third Set of Locks Project. Mr Kaufmann is on the Board of Directors for the Environmental and Engineering Geophysical Society and is a Section Officer for the Association of Environmental and Engineering Geologists.

Acquisition and Analysis of Terrestrial Gravity Data

LELAND TIMOTHY LONG
Professor Emeritus, Georgia Institute of Technology

RONALD DOUGLAS KAUFMANN
President, Spotlight Geophysical Services

CAMBRIDGE
UNIVERSITY PRESS

CAMBRIDGE UNIVERSITY PRESS
Cambridge, New York, Melbourne, Madrid, Cape Town,
Singapore, São Paulo, Delhi, Mexico City

Cambridge University Press
The Edinburgh Building, Cambridge CB2 8RU, UK

Published in the United States of America by Cambridge University Press, New York

www.cambridge.org
Information on this title: www.cambridge.org/9781107024137

First published 2013

Printed and bound in the United Kingdom by the MPG Books Group

A catalogue record for this publication is available from the British Library

Library of Congress Cataloguing in Publication data
Long, L. T. (Leland Timothy)
Acquisition and analysis of terrestrial gravity data / Leland Timothy Long, Professor Emeritus,
Georgia Institute of Technology; Ronald Douglas Kaufmann, Spotlight Geophysical Services.
pages   cm
Includes bibliographical references and index.
ISBN 978-1-107-02413-7
1. Gravity anomalies – Measurement.   2. Geophysical surveys.   3. Earth – Crust.
I. Kaufmann, Ronald Douglas.   II. Title.
QB337.L66   2013
526′.7 – dc23   2012030766

ISBN 978-1-107-02413-7 Hardback

Dedicated to our mentors and students.

In particular to Dr. Alan R. Sanford who instilled the importance of examining problems in revolting detail, to Dr. Joseph W. Berg for his enthusiasm for new ideas and participation in the science community, and to the many students who participated in the acquisition and analysis of gravity data while at Georgia Tech and who have continued these virtues in their work, teaching, and life.

Contents

Preface

Gravity data acquisition and analysis is most often presented in outline form as one of the smaller chapters in books on general geophysical exploration methods. This limited description means that the details of field techniques and data analysis are lost or greatly abbreviated and left to the individual to learn through experience. The objective of this book is to offer a detailed presentation of gravity data acquisition and analysis in a single package. The examples are taken from geophysical engineering problems as well as the analysis of regional and global data.

The objective is to completely cover the information needed for a novice to understand how and why gravity data are acquired and analyzed. A student completing a course using this text could easily acquire gravity data and would be prepared to initiate independent research on the analysis of potential data. A consulting geophysicist will find a base of both theoretical- and application-oriented information in this text, while a geologist or engineer can use this book to better understand the advantages and limitations of the gravity method. The general approach of this text has evolved over the past 20 years through experience gained from the acquisition of more than 40,000 values of gravity and in teaching courses in potential methods.

The text is intended for a wide range of users. It is written so that the basic applications are easily understood by those with limited training in mathematics. At the same time, the text occasionally introduces more advanced topics from potential theory for those with greater skills in mathematics. The text does not present extensive equations for the many possible specific models. Some simple shapes lead to complex equations that are computationally intense and generally of little practical use. Instead, the text presents the simpler models as a means of illustrating concepts or as a method of approximating structures. Sufficient background is presented in the equations and analysis techniques for those wishing to create their own more detailed models. In general, methods that allow automatic modeling of the gravity fields using approximations will be emphasized. Inversion methods are presented for the geophysicists needing more advanced analysis techniques for larger datasets.

Most texts are lacking on advice for data acquisition and the quality control needed to prevent corruption of the data. This text presents important aspects of data acquisition and provides organizational tools needed to carry out a successful survey. The text draws on examples from a wide variety of acquisition problems. They range from environmental and engineering problems, such as locating sink holes, to examples from crustal structure analysis and global gravity fields.

This book is a user's manual for those wishing to obtain and use gravity data, as well as a textbook for the introduction to more advanced concepts in tectonics and geodesy.

1 Gravitational attraction

1.1 Universal gravitational attraction

In his *Principia*, which he completed in 1686, Sir Isaac Newton demonstrated the inverse square law for universal gravitation. This law has provided the mathematical basis for the study of gravitational attraction among large attracting masses. Newton demonstrated that the inverse square proportionality for the attractive forces among all matter could be shown from his second law of motion and Kepler's third law of planetary motion. This gravitational force is one of the weakest forces in nature by many orders of magnitude. The sizes of the masses involved give it importance on Earth and in space. Partly because of its weakness, physicists today still argue about the role of gravity in relativity and particle physics, occasionally suggesting that at some scales the inverse square law may break down. However, for the study of the gravitational attraction of the Earth and its geological features, Newton's inverse square law and the larger field of potential theory derived from it are sufficient.

Newton's second law relates the force acting on a body to its change in momentum. In order to demonstrate the inverse square law, one can consider a point mass, m, or planet, in orbit about a larger mass, m'. The point mass experiences an attractive force pulling it back toward the larger mass (Figure 1.1). On traveling a short arc distance, c, the planet is pulled a distance, s, from its straight line path to a position closer to m'. From Newton's second law the distance, s, which is the distance a body would move under a constant acceleration, is given by the relation

$$s = \frac{1}{2}a(\delta t)^2, \tag{1.1}$$

where a is the acceleration experienced by the planet and δt is the time increment the planet takes to travel the distance c. The time increment can be eliminated by equating its fraction of the total period of revolution, T, to the ratio of the small arc distance, c, to the circumference of the orbit

$$\frac{\delta t}{T} = \frac{c}{2\pi R}. \tag{1.2}$$

Equation (1.2) allows the acceleration to be written in the form

$$\frac{a}{2} = s\left(\frac{2\pi R}{Tc}\right)^2. \tag{1.3}$$

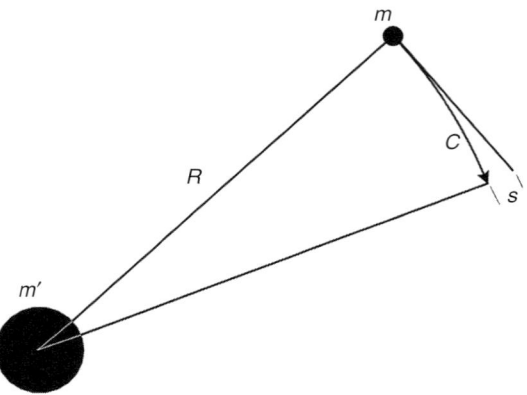

Figure 1.1   Path of particle *m* in orbit about a large planet of mass *m'*.

By a geometrical proof, Newton showed that $c^2 = 2Rs$ for arc distances c much less than the radius, R, and was able to eliminate both s and c^2 in Eq. (1.3). The resulting relation contains the square of the period of revolution. Kepler's third law, which is the observation that the squares of the periods of any two planets are proportional to the cubes of their mean distances from the Sun, provided the relation needed to demonstrate the inverse square law of universal gravitational attraction. The proportionality takes the form

$$a = 2s \left(\frac{2\pi R}{Tc} \right)^2 \geq \frac{4\pi^2 R}{T^2} \approx \frac{1}{R^2}. \tag{1.4}$$

The constant of proportionality turns out to be proportional to the attracting mass, m', and, therefore, can be written Gm', where G is the universal gravitational constant. The force of attraction between the two masses is therefore

$$F = -G \frac{m'm}{R^2}, \tag{1.5}$$

where:

    $F = ma$, the magnitude of the force of attraction
    $G =$ the universal gravitational constant
    $m' =$ the attracting mass of the Sun
    $m =$ the attracted mass of the planet
    $R =$ the distance from the center of m to the center of m'

The negative sign designates the force as attractive, toward the center of m'.

The gravitational forces on m and m' are equal in magnitude, opposite in direction, and along the line joining the two masses. The force is proportional to the product of the two masses and inversely proportional to the square of their separation. The value of G based on early pendulum measurements was $(6.67428 \pm 0.00067) \times 10^{-11}$ m^3 kg^{-1} s^{-2}. The best measurements of G as of 2010 (Mohr, et al., 2011) is $(6.67384 \pm 0.00080) \times 10^{-11}$ m^3 kg^{-1} s^{-2}.

For more than 300 years, Newton's law has provided the basis for studies of gravitational attraction, a remarkable record for a relation based on empirical observations. Recent

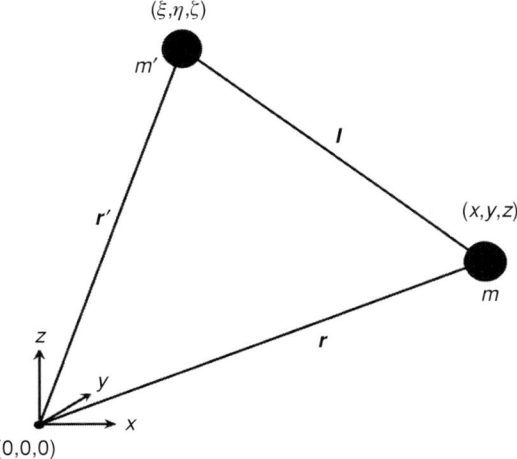

Figure 1.2   Vector notations for attraction of point masses.

speculations about a fifth force suggest a deviation from the inverse square relation, but these have proven too small, if they even exist, to be confirmed by measurements. Also, Einstein's special relativity added a new dimension to Newtonian space-time that has changed how physicists perceive gravitation. However, the relativistic deviations from Newtonian gravitation are too small to have an impact on the measurement of the Earth's gravitational attraction because the velocities in the Earth's system are insignificant relative to the speed of light. These relativistic effects have only recently become measurable in satellite data.

1.2 Gravitational acceleration

The acceleration of a small point mass near a much larger attracting mass, such as would approximate the attraction of satellites orbiting the Earth in space, can be expressed as the acceleration of a unit mass, $m = 1.0$. The magnitude, a, of the acceleration of a unit mass is

$$a = \frac{F}{m} = -G\frac{m'}{r^2}. \tag{1.6}$$

Equation (1.6) gives the gravitational attraction of the larger mass m'. Because the force of attraction and acceleration are vectors, a general expression with the origin displaced from the position of the attracting mass is more appropriate and is needed for more complex computations. The direction vector, $\boldsymbol{l}$, along which the force acts, is the difference in the position vectors $\boldsymbol{r}$ and $\boldsymbol{r'}$ of the two masses (Figure 1.2). The position vectors $\boldsymbol{r}$ and $\boldsymbol{r'}$ point from the origin to m and m', respectively. The gravitational attraction in vector notation may be expressed as

$$\boldsymbol{a} = -G\frac{m'(\boldsymbol{r} - \boldsymbol{r'})}{|\boldsymbol{r} - \boldsymbol{r'}|^3} = -G\frac{m'}{|\boldsymbol{l}|^2}\frac{\boldsymbol{l}}{|\boldsymbol{l}|}. \tag{1.7}$$

In Cartesian coordinates, the attracted mass, usually the unit test mass, m, is at the position (x, y, z), and the attracting mass is at position (ξ, η, ζ). In applications to the Earth, the origin is the center of mass of the Earth and the z-axis is the mean axis of rotation. The x- and y-axes are arbitrary, but x is by convention the meridian plane of Greenwich, England. This orientation of the reference axes defines the geocentric coordinate system for the Earth. Expanding Eq. (1.7) in terms of the coordinates of m and m' gives

$$[a_x, a_y, a_z] = -G \frac{m'\,[x - \xi, y - \eta, z - \zeta]}{\left((x - \xi)^2 + (y - \eta)^2 + (z - \zeta)^2\right)^{3/2}}, \tag{1.8}$$

where in Eq. (1.8), $|l|$ has been replaced by $\sqrt{(x - \xi)^2 + (y - \eta)^2 + (z - \zeta)^2}$. The components of the vector, $\boldsymbol{a}$ in the directions of the geocentric coordinated axes are $[a_x, a_y, a_z]$.

Equation (1.8) can be modified and simplified for computation of the effects of anomalous mass near the Earth's surface. The attraction of the anomalous mass is typically less than 0.001 percent of the attraction of the Earth. Also, the curvature of the Earth's surface may be neglected for all but the larger regional surveys. Given the small magnitude of local anomalies and negligible difference between a flat plane and the survey area for local surveys, a rectangular coordinate system can be used. The vertical direction is set to coincide with the direction of the Earth's gravity field, the direction in which measurements of the magnitude of the gravity field are determined. The attraction of anomalous mass is projected onto the vertical direction and two horizontal directions (typically north and east) for computation. In most gravity measurements for the interpretation of anomalous density structures near the Earth's surface, the main portion of the Earth's field is removed and only the anomalous component is retained. In Eq. (1.8) the vertical direction corresponds to the radial or normal direction. For application to the Earth's surface in local surveys, the radial direction is usually assigned the z direction in a rectangular coordinate system. The attraction of anomalous masses at the surface, from Eq. (1.8) is

$$a_z = -G \frac{m'(z - \zeta)}{\left((x - \xi)^2 + (y - \eta)^2 + (z - \zeta)^2\right)^{3/2}}, \tag{1.9}$$

where the z-axis now refers to the vertical direction which is in line with the negative direction of the Earth's gravity field. The ratio of $z - \zeta$ to l is the cosine of the angle between vertical (the z-axis) and the attraction of anomalous mass, and thus for anomalous mass, Eq. (1.9) is the projection of the attraction of anomalous mass onto the z-, or vertical, axis.

1.3 Gravitational potential of a point mass

The gravitational vector field can be derived from a potential scalar field because the gravitational field is conservative. For a conservative vector field, the work required to move a particle from point A to B is independent of the path (Figure 1.3). The work

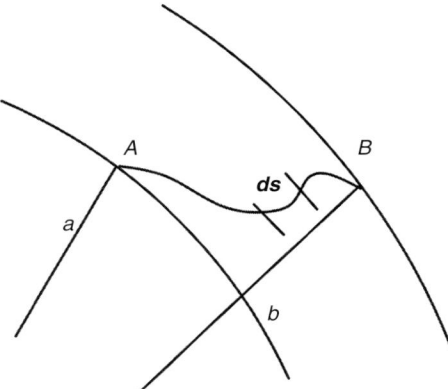

Figure 1.3   Work along path from point A to B. A and B are at different potential levels.

required, in the absence of friction, is the integral along the path of the product of force times the distance moved,

$$\Delta W = W(B) - W(A) = \int_A^B \boldsymbol{F} \cdot \boldsymbol{ds},\tag{1.10}$$

where the dot product gives the component of the force in the direction of movement given by $\boldsymbol{ds}$. If we equate a for a point mass from Eq. (1.7) to $\boldsymbol{F}$, Eq. (1.10) becomes,

$$\Delta W = -G\int_A^B \frac{m'\boldsymbol{l}\cdot\boldsymbol{ds}}{l^3} = -G\int_A^B \frac{m'l\cos(\theta)\,ds}{l^3} = -G\int_{l_A}^{l_B} \frac{m'dl}{l^2} = Gm'\left(\frac{1}{l_A} - \frac{1}{l_B}\right),\tag{1.11}$$

where dl is $l\cos(\theta)ds$ the projection of $\boldsymbol{ds}$ on $\boldsymbol{l}$. The gravitational potential V is the limit of ΔW as l_A goes to infinity. The expression for the potential is

$$V = -G\frac{m'}{l}.\tag{1.12}$$

In practice, the solutions for many gravity problems are easier to solve by using the scalar potential and computing the gravitational acceleration from the gradient of the potential

$$\boldsymbol{a} = -\nabla V.\tag{1.13}$$

1.4 Gravitational potential of a solid body

The gravitational attraction of a composite body, such as the Earth, is the combination of the attractions of its countless mass elements. The total attraction is the vector summation of the accelerations of the individual elements or the gradient of the scalar summation of the potentials for all the mass elements. The expression for the potential for many point masses can be written as the sum

$$V = G\frac{m_1}{l_1} + G\frac{m_2}{l_2} + G\frac{m_3}{l_3} + \cdots. \tag{1.14}$$

By letting the mass elements become smaller and more numerous, the potential for an anomalous distribution of mass can be approximated to any degree of precision. In the limit of infinitesimally small mass increments the summation can be replaced by the integral

$$V = G\int \frac{dm}{l}, \tag{1.15}$$

where dm is an infinitesimally small mass increment. The density ρ of the medium is defined as the ratio of the mass to the volume of the mass in the limit as the volume goes to zero,

$$\rho = \lim_{\delta v \to 0} \frac{m}{v}. \tag{1.16}$$

Density is a scalar function of position in an anomalous mass. The density distribution can be highly discontinuous and irregular in real materials, and may vary radically across grain boundaries, in voids, in caverns, and at the surface where rock comes in contact with the atmosphere. In practice, average or smoothed values of the density distribution are used in computation. By substituting the expression for dm in terms of density into Eq. (1.15), the integral expression for the potential is an integral over the volume,

$$V = G\int_v \frac{dm}{l} = G\int_v \frac{\rho dv}{l}. \tag{1.17}$$

In rectangular coordinates the potential for a solid body from Eq. (1.17) is written as

$$V(x, y, z) = G\int_v \frac{\rho(\xi, \eta, \zeta)d\xi d\eta d\zeta}{\left\{(x-\xi)^2 + (y-\eta)^2 + (z-\zeta)^2\right\}^{1/2}}, \tag{1.18}$$

where $dv = d\xi d\eta d\zeta$.

The gravitational acceleration in the three orthogonal coordinate directions, can be found by differentiating Eq. (1.18) by x, y, and z, giving, respectively,

$$a_x = \frac{\partial V}{\partial x} = -G\iiint \frac{\rho(\xi, \eta, \zeta)(x-\xi)d\xi d\eta d\zeta}{\left\{(x-\xi)^2 + (y-\eta)^2 + (z-\zeta)^2\right\}^{3/2}} \tag{1.19}$$

$$a_y = \frac{\partial V}{\partial y} = -G\iiint \frac{\rho(\xi, \eta, \zeta)(y-\eta)d\xi d\eta d\zeta}{\left\{(x-\xi)^2 + (y-\eta)^2 + (z-\zeta)^2\right\}^{3/2}} \tag{1.20}$$

$$a_z = \frac{\partial V}{\partial z} = -G\iiint \frac{\rho(\xi, \eta, \zeta)(z-\zeta)d\xi d\eta d\zeta}{\left\{(x-\xi)^2 + (y-\eta)^2 + (z-\zeta)^2\right\}^{3/2}}. \tag{1.21}$$

In computing the gravitational attraction, it is often easier to integrate the potential once and differentiate than it is to solve the integration three times.

The attraction between two static rigid bodies, each too irregular in shape to assume they are equivalent to a point mass, requires integrations over the total volume of both masses. For example, the gravitational attraction in the x direction, for the center of mass of two irregular bodies is given by the six integrations

$$a_x = G \iiint\limits_{\xi_1\ \eta_1\ \zeta_1}\ \iiint\limits_{\xi_2\ \eta_2\ \zeta_2} \frac{\rho_1\,(\xi_1, \eta_1, \zeta_1)\,\rho_2\,(\xi_2, \eta_2, \zeta_2)\,(\xi_1 - \xi_2)\,d\xi_1 d\eta_1 d\zeta_1 d\xi_2 d\eta_2 d\zeta_2}{\left\{(\xi_1 - \xi_2)^2 + (\eta_1 - \eta_2)^2 + (\zeta_1 - \zeta_2)^2\right\}^{3/2}}.$$

$$(1.22)$$

For moving systems the asymmetry of mass can lead to rotational forces, such as those exerted by the Sun and Moon on the bulge of the rotating Earth. These forces create a torque that explains the precession of the Earth's axis of rotation.

1.5 Surface potential

When the distribution of mass is restricted to a thin two-dimensional sheet, it is convenient to express the equation for the potential as a two-dimensional integral over that surface. For computation, the thickness is assumed to go to zero and the sheet has a mass density of $\kappa = dm/ds$, defined by the ratio of mass, dm, to surface area, ds. In practice, the sheet only needs to be thin relative to the distance of the attracted point for the computation to be useful in modeling. In this case the surface density is the limit as the thickness goes to zero of the product of density, ρ, and thickness, dh,

$$\kappa = \frac{dm}{ds} \cong \rho dh. \tag{1.23}$$

The integral for the potential can be expressed in various forms, such as

$$V = G \iiint\limits_{v} \frac{dm}{l} = G \iiint\limits_{v} \frac{\rho dv}{l} = G \iiint\limits_{v} \frac{\rho dh ds}{l} = G \iiint\limits_{v} \frac{\kappa ds}{l}. \tag{1.24}$$

On the surface, the normal derivatives are discontinuous. They differ according to whether the derivative is taken from the internal or external side of the surface,

$$\left.\frac{\partial V}{\partial n}\right|_e = -2\pi G\kappa + G \iint\limits_{\sigma} \kappa \frac{\partial}{\partial n}\left(\frac{1}{l}\right) d\sigma \tag{1.25}$$

$$\left.\frac{\partial V}{\partial n}\right|_i = +2\pi G\kappa + G \iint\limits_{\sigma} \kappa \frac{\partial}{\partial n}\left(\frac{1}{l}\right) d\sigma \tag{1.26}$$

$$\left.\frac{\partial V}{\partial n}\right|_i - \left.\frac{\partial V}{\partial n}\right|_e = 4\pi G\kappa. \tag{1.27}$$

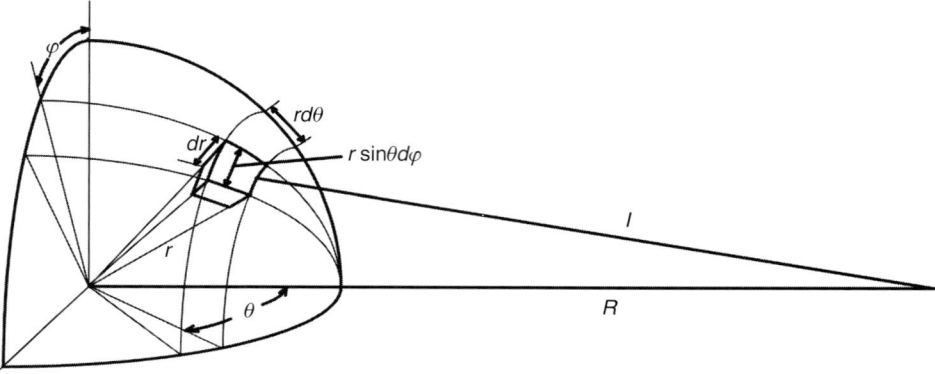

Coordinate system for computation of the potential for a constant density sphere.

1.6 Attraction of a sphere

The attraction of a sphere of uniform density is, perhaps, the most useful and fundamental relation for the interpretation of gravity anomalies. It is a first approximation to the attraction of any compact irregularly shaped body of mass at distances that are greater than the diameter of the body. In order to demonstrate the integration, Eq. (1.18) is expressed in spherical coordinates in which the incremental volume is $r^2 \sin\theta \, dr \, d\theta \, d\phi$, where the coordinates are defined in Figure 1.4. The expression for the potential in spherical coordinates is

$$V = G\rho V = G\rho \int_0^a \int_0^\pi \int_0^{2\pi} \frac{r^2 \sin\theta}{l} dr \, d\theta \, d\phi. \tag{1.28}$$

The integration over the shell is computed first. The integration over the longitudinal coordinate, λ, is trivial because the distance, l does not change with a change in ϕ, and the integration simply gives the factor of 2π,

$$V = 2\pi G\rho \int_0^a \int_0^\pi \frac{r^2 \sin\theta}{l} dr \, d\theta. \tag{1.29}$$

The distance l is a function of a, but it can be expressed in spherical coordinates through the law of cosines for the triangle formed by the origin, the attracted mass and the incremental attracting mass

$$l^2 = r^2 + R^2 + 2rR\cos\theta. \tag{1.30}$$

By differentiating Eq. (1.30) with respect to θ it can be shown that

$$\frac{dl}{rR} = \frac{\sin\theta}{l} d\theta. \tag{1.31}$$

This relation may be substituted back into Eq. (1.29) to change the variable of integration,

$$V = 2\pi G\rho \int\limits_{0}^{a} \int\limits_{R-r}^{R+r} \frac{r^2}{rR} dr\,dl. \tag{1.32}$$

Evaluation of the integral for l gives

$$V = 2\pi G\rho \int\limits_{0}^{a} \frac{r^2 dr}{rR} l\Big|_{R-r}^{R+r} = 2\pi G\rho \int\limits_{0}^{a} \frac{r^2 dr}{rR} [(R+r) - (R-r)], \tag{1.33}$$

or

$$V = -4\pi G\rho \int\limits_{0}^{a} \frac{r^2 dr}{R} = -\frac{4\pi}{3} a^3 G\rho \frac{1}{R}. \tag{1.34}$$

The force of attraction of the sphere in the radial direction is the derivative with respect to R, or

$$a_x = -\frac{\partial V}{dR} = \frac{4\pi}{3} a^3 G\rho \frac{2}{R^2}. \tag{1.35}$$

This radial direction is referenced to the center of mass of the sphere. The integration has demonstrated that the attraction of a sphere is equivalent to the attraction from a point at the center of the sphere with all the mass concentrated at the center.

For a small spherical zone of anomalous density near the Earth's surface, the anomalous field due to the sphere will be proportional to the density contrast between the sphere and the Earth. Because the Earth's gravity field is on the order of 6 orders of magnitude greater than anomalous fields, the anomalous attraction of a spherical shaped anomaly can be measured in the vertical direction. As in Eq. (1.21), the vertical component of the attraction of a sphere is given by the derivative of Eq. (1.34) with respect to the vertical, z,

$$a_z = \frac{\partial V}{\partial z} = \frac{4\pi}{3} a^3 G\Delta\rho \frac{(z-\zeta)}{\left[(x-\xi)^2 + (y-\eta)^2 + (z-\zeta)^2\right]^{3/2}}. \tag{1.36}$$

1.7 Units of acceleration

The S.I. units of g are m/s^2, although other units are still frequently used. The practical unit for measurements of variations in the Earth's gravity is on the order of μm/s^2. One μm/s^2 corresponds to the "gravity unit" or "g.u." originally used in oil exploration geophysics. Although S.I. units are preferred or required by most journals, some recent literature in the field of geophysics still uses the Gal, one cm/s^2, for presentations of gravity data. In these publications the mGal $= 0.001$Gal $= 10$ μm/s^2 is the most common unit for contouring gravity data.

2 Instruments and data reduction

2.1 The gravitational constant

The gravitational attraction of the Earth according to Newton's universal law of gravitation is proportional to the mass of the Earth and inversely proportional to the distance from the Earth's center. The constant of proportionality is the universal gravitational constant, G. In measuring gravitational attraction the mass of any planetary body and G are coupled and planetary observations cannot be used to determine the independent values of G and mass. Independent measurements of G and the mass of the Earth, or equivalently its mean density, have not been easy. The value of G is currently defined to four significant figures,

$$G = (6.67384 \pm 0.00080) \times 10^{-11} \text{ m}^3 \text{ kg}^{-1} \text{ s}^{-2}, \qquad (2.1)$$

or 10^{-8} dyne cm^2/g^2. This most recent value of G is from the 2010 CODATA Recommended Values for Physical Constants, which were released June 2011 by the National Institute of Standards and Technology. Physicists still argue about the meaning of gravity and whether G is truly a constant; but, for the practical study and measurement of the Earth's shape and its structures, the implications of these arguments are insignificant.

The first experiment capable of determining the universal gravitational constant was carried out in 1798 by Cavendish. The Cavendish apparatus, a torsion balance, made use of the attraction of spheres, where the attraction of a sphere is known to be the same as the attraction of a point mass at the center of the sphere. A large mass, M in Figure 2.1, is moved into position and the deflection, ε, of a torsion balance is observed. The small mass and elastic constant can be expressed in terms of the period of the torsion balance without the large mass. Using the ratio of the size of the two masses and the attraction of one mass to the Earth, Cavendish was able to compute the Earth's mean density. Actual computations of G came much later from similar measurements. The more accurate recent measurements determine the period with and without the larger mass. The improvement in the recent measurements comes from improvements in the ability to measure period more accurately and from the increased precision in displacement afforded with laser interferometers. Parks and Faller (2010) used a laser interferometer to measure the change in spacing between two free-hanging pendulum masses to provide one of the most recent and precise measurements of G.

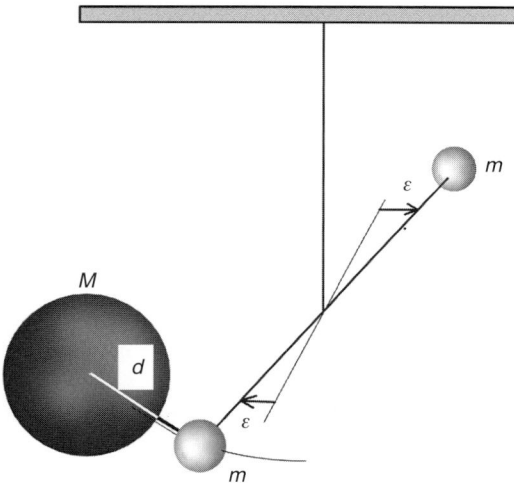

Figure 2.1   Schematic illustration of the Cavendish torsion balance.

2.2 Absolute measurements

An "absolute" gravity measurement is the determination of the gravitational acceleration from measurements of displacement or time. Pendulum measurements were used first to determine the value of the gravitational attraction of the Earth. The pendulum methods are based on the measurement of the period of a freely swinging pendulum. Although the period can be measured with considerable precision, the accuracy of pendulum measurements depends on careful measurement of the dimensions of the pendulum and its dynamic interaction with the pendulum support. Other sources of error include elastic deformation of the components of the pendulum and the rolling of a knife edge often used to reduce friction. For a point mass suspended on a weightless inelastic wire, a mathematical analog for a pendulum shown in Figure 2.2, the period, T, is given by

$$T = 2\pi \sqrt{\frac{l}{g}} \left(1 + \frac{\sin^2 \frac{\varepsilon}{2}}{4} + \frac{9}{64} \sin^4 \frac{\varepsilon}{2} \cdots \right), \tag{2.2}$$

where l is the length of the pendulum and ε is the maximum swing of the pendulum. For a maximum swing of two degrees, the period is 1.000076 times the zero swing length. For a maximum swing of 60 degrees, the factor is 1.07.

The inherent inaccuracy in determining the effective length was overcome by the design of a reversible pendulum. A series of precise measurements of the acceleration of gravity at Potsdam, Germany, were made using the reversible pendulum in 1906. The resulting value of 981274 ± 3 mGal was adopted in the 1930s as the international standard for measurement of gravity data worldwide. The Potsdam, Germany, value was tied to national and regional (State level) networks. The station locations were typically at airports and court

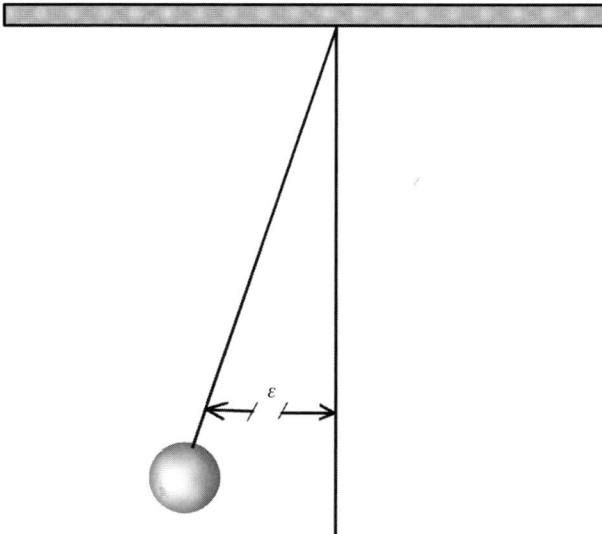

Schematic illustration of the simple pendulum.

houses because it was thought at the time that access would be quick and easy and that these facilities would remain over time. Absolute gravity measurements were subsequently made at many sites around the world. The combined comparison of these independent absolute measurements with ties to Potsdam demonstrated that the 1906 Potsdam measurement was consistently too high. The value stood until a re-examination of the correction for elastic response decreased the value by about 15 mGal. In 1971, after a massive examination of absolute and relative gravity measurements worldwide, a new standard for international measurements was adopted and referred to as the 1971 International Gravity Standardization Net (IGSN71) (Morelli, 1974). The concept of the IGSN 71 differs from that of the Potsdam system in that the gravity datum is determined, not by an adopted value at a single station, but by a least squares adjustment of 1854 stations with respect to ten absolute gravity values at eight stations in North America, South America, and Europe

More recently, additional corrections to the IGSN71 datum have been proposed (Hinze et al., 2005). Because regional surveys are often referenced to an absolute gravity datum (such as the IGSN71), it is important that the same gravity datum be used when combining data from multiple surveys. Caution is advised in combining old and recent data sets.

The development of distance-measuring equipment using laser interferometer in the 1980s made field measurements of absolute gravity possible, although at a much higher cost and longer acquisition time than comparable measurements made with relative gravity meters. The operation of a ballistic absolute gravity meter is based on the measurement of the acceleration of a falling body. The equation for free fall in a vacuum, isolated from the viscous drag of air, is the well-known relation

$$x = x_0 + v_0 t + \frac{g}{2} t^2,$$ (2.3)

where x_0 and v_0 are, respectively, the initial position and initial velocity. With three pairs of measurements of x and t, the initial position and velocity can be eliminated and a value of g obtained from

$$g = 2\frac{(x_3 - x_1)(t_2 - t_1) - (x_2 - x_1)(t_3 - t_1)}{(t_3 - t_1)(t_2 - t_1)(t_3 - t_2)}. \tag{2.4}$$

The accuracy of absolute gravity measurements is now better than 0.01 mGal. However, a ballistic measurement of acceleration gives the instantaneous value of acceleration, which includes the motion of the ground related to background seismic vibrations, microseisms, and tidal displacements. Because this precision is the same order of magnitude as the acceleration of the ground from seismic noise, many such measurements are averaged or correlated with seismic noise measurements in order to improve precision. Networks of base stations with gravity determined by these instruments are replacing the existing standard for international gravity measurements (IGSN71).

2.3 Relative measurements

Most gravity measurements used in geophysical surveys are relative measurements. Relative measurements are made by instruments that can compare the gravitational attraction at a field site with the attraction at a base station where the absolute value of gravity is known or assumed. Some local surveys may not have access to a calibrated base station or need absolute gravity values for comparison with other surveys. Because a comparison of measurements is being made, relative measurements do not require the same degree of care in determining the instrument's dimensions and elastic response that have been problematic for absolute gravity measurements.

A portable pendulum was commonly used for field measurements in the first half of the twentieth century. In a pendulum measurement system, the term for length is replaced by the expression for gravity at the site where gravity is known and the change in period can be related directly to the change in gravity by

$$\Delta g_{12} = g_2 - g_1 = 2g_1\frac{T_1 - T_2}{T_1} + \cdots. \tag{2.5}$$

Thus, the length measurement is not required, as it is in absolute measurements. However, the time required for setting up and recording a gravity value was on the order of a day and the stability of these systems was limited by the stability and resonance of the measurement platform. The precision of the pendulum system was on the order of 1 mGal, which is not sufficiently precise by today's standards.

Most relative gravimeters today utilize a spring balance. In principle, the gravitational force, mg, of a small mass is balanced by the restoring force of a spring, kx, where k is the equivalent of a Young's modulus for the spring. The relation for a simple linear spring balance,

$$mg - kx = 0, \tag{2.6}$$

requires unreasonable precision in measuring x, the displacement. A more general expression for a spring balance system may be written as

$$gM(z, T, P, \alpha, \cdots) + \overline{M}(z, T) = 0, \tag{2.7}$$

where M and $\overline{M}$ are the gravitational and internal restoring moments, respectively. The moments are generally functions of the parameters, z, T, P, α; respectively, distortion, temperature, pressure, and tilt. These are the principal components, but there could be other factors that would cause a variation in the gravitational moment. The sensitivity of a meter can be computed from the differential with respect to g of the more general expression for a spring balance. The expanded expression is

$$M\frac{dg}{dg} + \left(g\frac{\partial M}{\partial z} + \frac{\partial \overline{M}}{\partial z}\right)\frac{dz}{dg} + \left(g\frac{\partial M}{\partial T} + \frac{\partial \overline{M}}{\partial T}\right)\frac{dT}{dg} + g\frac{\partial M}{\partial P}\frac{dP}{dg} + g\frac{\partial M}{\partial \alpha}\frac{d\alpha}{dg} = 0. \tag{2.8}$$

In most modern sensitive instruments, the temperature T is held constant by an internal heater and the tilt α is held constant by leveling. Pressure P generally does not vary significantly or is eliminated by placing the meter in a sealed chamber. If the temperature, pressure, and tilt are unchanged, then the sensitivity reduces to

$$\frac{dg}{dz} = \frac{-M}{\left(g\dfrac{\partial M}{\partial z} + \dfrac{\partial \overline{M}}{\partial z}\right)} = \frac{M}{\phi(z)}. \tag{2.9}$$

In Eq. (2.9) the sensitivity can be made very large if the denominator can be made close to zero, allowing a very precise measure of the relative gravitational acceleration. The object is to design the internal workings of a meter so that the internal restoring moment is negative and cancels out the gravitational moment. Such systems are referred to as astatic and are inherently non-linear. The objective is to make the negative restoring moment as close to the gravitational moment as possible. If it exceeds the gravitational moment, the sign in the denominator changes and the meter becomes unstable. The high sensitivity and finite dimensions of these systems require a non-linear response for large deflections. Hence, the high sensitivity can be achieved only in a narrow measurement range, the null position, or the measuring line. A gravity measurement is made by measuring the force required to return the mass to the measuring line, or null position. At the measuring line, all parameters are the same, except for the gravitational acceleration and restoring force.

A simple example of an astatic device is the inverted pendulum. As illustrated in Figure 2.3 when the mass moves proportionally further off the vertical, a destabilizing torque about the axis increases in proportion to $\sin\theta$, where θ is the angle of deflection away from the equilibrium position. The sensitivity is the difference between the spring restoring force and the gravitational pull of the mass of the inverted pendulum. The weight of the small mass is then proportional to the angle of deflection.

Of the many types of gravity meters available, the most common in the late 1900s were the *Lacoste–Romberg* and the *Worden*. The Worden uses a horizontal beam and a spring assembly made of quartz glass packaged in a thermally insulated vacuum. The

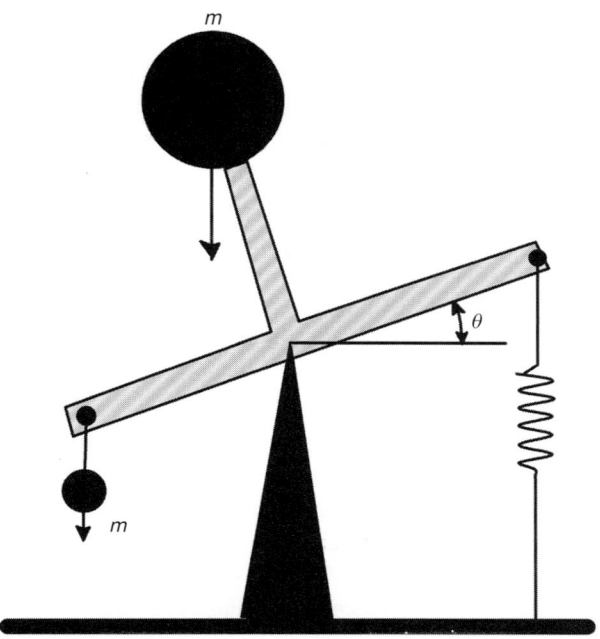

Figure 2.3   Schematic design of an astatic device utilizing the inverted pendulum.

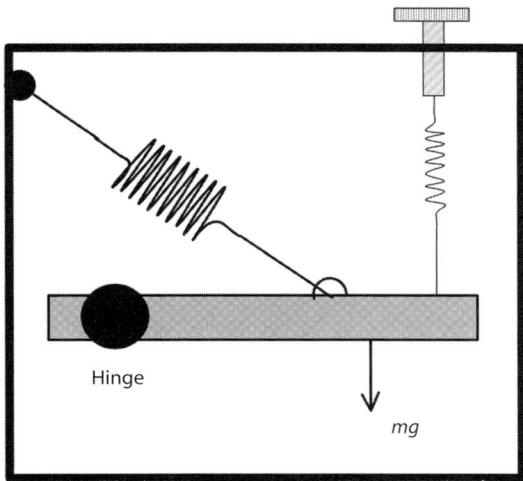

Figure 2.4   Schematic design for the Lacoste–Romberg and North American gravity meters.

Lacoste–Romberg uses a zero-length iso-elastic metal spring in an arrangement illustrated in Figure 2.4. The iso-elastic properties reduce the effects of deformation of the spring under load. Many materials used in common springs that are not iso-elastic have a tendency to change spring constant under stress as portions of the crystalline structure realign. The zero length in the Lacoste–Romberg geometry generates a destabilizing force that is canceled at small angles by the change in force associated with the change in length as the mass rotates about its hinge. The result is a narrow measurement zone with a restoring force that can be adjusted arbitrarily small.

More modern gravity meters have adapted electronic data acquisition features, which include digital signal enhancement options for improved accuracy and speed of measurement as well as integrated GPS location and precise timing. These features have been added to older meters such as the Lacoste–Romberg Model D and G meters and integrated into new meters, such as the Scintrex CG-5 gravimeter and the ZLS Corporation Burris gravimeter. The Scintrex CG-5 improves upon the entirely mechanical operation of older meters by using a fused quartz spring with electrostatic nulling to provide a measurement range of 8000 mGal with a reading resolution of 0.001 mGal.

The measurement line, which is the position of maximum sensitivity, can be verified by systematically tilting the meter equal amounts in alternate directions perpendicular to the spring direction. The center line is the position that gives an equal displacement in both directions. The sensitivity can be adjusted by tilting the meter parallel to the spring. The level indicators on a gravity meter would normally be pre-set for optimal sensitivity. Tilt in the direction of the hinge of the mass will increase the period. The period is typically set to approximately 20 s. Because the motion of the mass is damped, the meter will take approximately 20 s to reach equilibrium. The damping and long period effectively eliminate short-period vibrations and the automatic signal stacking and digital filtering available on newer gravity meters reduces the effects of these noise sources on the final data. However, the effects of the longer-period motions of surface waves from major earthquakes and microseisms from ocean swells are more difficult to remove.

2.4 Instruments for gravity-gradient measurements

Instruments are now available to directly measure the gravity gradient, the spatial rate of change of gravitational attraction. There are advantages and disadvantages to the use of instruments that directly measure the various components of the gravity gradient. The principal advantage is the removal of a datum and suppression of long-wavelength anomalies. Also, the gravity gradient can be used to measure the density of rocks down a well. The enhancement of short-wavelength anomalies is both an advantage and disadvantage. Anomalies appear sharper when viewed as a gradient anomaly, but the noise in short-wavelength data will be amplified significantly. Also, because gravity-gradient measurements are often made in moving vehicles, navigation errors and the influence of high accelerations will deteriorate the precision, an effect that is of considerable importance because gravity-gradient measuring instruments require more precise and complex design than the traditional relative-gravity instruments. The gravity gradient of an anomaly is discussed in connection with the use of filters to manipulate gravity data. The computation of gravity-gradient anomalies will not be expounded on in this text because the gradient anomaly can be computed directly by taking the difference of two close values of the gravity anomaly, or by simple modification of the basic equations.

The unit of gravity gradient is the Eötvös, or E, which is equivalent to 10^{-9} s^{-2} or 10^{-4} mGal/m. Conventional gravity measurements are the acceleration of the Earth and are measured with the instrument oriented parallel to the direction of attraction, commonly referred to as the vertical. Measurements made at different heights give the vertical gradient, the change of the vertical component of gravity with respect to elevation. Measurements at different locations can give the horizontal gradients. The gradients are obtained numerically from variations in the values of gravity at different locations. Gravity gradiometer systems are rarely maintained in a vertical orientation and they typically measure the gradient in a direction that is not necessarily parallel to the vertical. The gravity gradient tensor is a nine-component symmetric tensor that defines the rate of change of the three components of the gravity vector. The tensor can be written as

$$
\nabla g = \begin{pmatrix}
\dfrac{\partial g_x}{\partial x} & \dfrac{\partial g_x}{\partial y} & \dfrac{\partial g_x}{\partial z} \\[2mm]
\dfrac{\partial g_y}{\partial x} & \dfrac{\partial g_y}{\partial y} & \dfrac{\partial g_y}{\partial z} \\[2mm]
\dfrac{\partial g_z}{\partial x} & \dfrac{\partial g_z}{\partial y} & \dfrac{\partial g_z}{\partial z}
\end{pmatrix}. \tag{2.10}
$$

A typical gravity gradiometer measures a single component of the tensor and the three components are measured using three orthogonal identical gradiometers. Given the orientation of the three instruments, the tensor can be rotated into the conventional vertical and two horizontal components of the gravity gradient.

Gravity gradiometer development has advanced rapidly with solid-state electronics and laser-aided measurements. The Lockheed Martin Gravity Gradiometers, which were originally developed for military applications, consist of two matched pairs of solid-state accelerometers mounted symmetrically around the edges of a rotating disk. The sum of the four signals is proportional to the gravity gradient over a distance proportional to the diameter of the disk. A single disk gives the gradient in one direction. In this arrangement, accelerations of the gradiometer cancel out, making it ideal for deployment in moving vehicles. The canceling out of accelerations common to all parts of the gradiometer is a critical component of all designs. A variety of accelerometers have been employed in the construction of gradiometers, including: electrostatic servo-controlled accelerometers, ribbons that respond to the gravity gradient, atom interferometers, and pairs of micro-flexure supported balance beams.

2.5 Data reduction

The usual target of a gravity survey is the variation in the gravity field that can be attributed to anomalous rock densities under a survey area. Unfortunately, variations in gravitational attraction caused by position in latitude and elevation are of the same magnitude or greater than those caused by anomalous densities. Also, the influence of topography in areas of high relief can distort the gravity field. Consequently, in order to reveal the signature in the gravity field caused by anomalous density structures, the influence of elevation, latitude, and topography must be removed.

In reducing gravity data, the first task is to remove those components of the gravity field and meter readings that change with time. Instrumental drift is a time-dependent change in the meter readings independent of actual changes in gravitational attraction. It is an inherent problem with relative gravity meters, but not with absolute meters. Drift in the meter reading is generally attributed to inelastic deformation of the elements of the meter under stress. Changes in the temperature of the meter, if not completely stabilized, can impact the drift of a meter reading through thermal expansion or contraction of elements in the meter. The characteristics and magnitudes of meter drift are unique for each individual meter, but usually gradual and linear enough to remove by periodically reoccupying a base station. With some meters, sudden changes in spring length can be caused by micro-fractures in the material or stress-induced changes in orientation of crystal zone of the spring metal. These sudden changes in the meter reading are called tares and can only be removed by reoccupying those stations visited before and after the tare.

The largest components of the natural variations in the gravitational field are caused by Earth tides with variations of up to 0.3 mGal within a day. The tidal variations are strictly periodic and can be predicted. Barometric pressure changes affect the gravity field to a much lesser extent, typically less than 0.003 mGal, per day, and depend on changes in the weather conditions. Over longer periods of time, changes in ground water level, subsidence or uplift, and construction adjacent to the site may cause the attraction of gravity to change between successive surveys. Over shorter periods of time, the shaking of the ground surface in an earthquake can make reading difficult.

The systematic variations in gravity of the lunar and solar tides are usually removed first. The remaining temporal variations are primarily related to meter drift, which typically is linear. The tidal gravity variations result from two causes; directly from the attraction of the Moon and Sun, and indirectly from the elastic deformation of the ground surface caused by these attractions and loading of the crust by ocean tides. The equilibrium theory of the tides, one of the many contributions from Newton's *Principia*, provides the basis for computing the tidal variations. The vertical tidal deformation at the Earth's surface is on the order of 0.2 m. This vertical deformation increases the equilibrium tides by a factor of around 1.2. Minor variations in this factor can be related to crustal structure, the elastic response of the crust, and crustal loading by ocean tides. Hence, theoretical computations of tides are not perfect and the difference can be used to study crustal elasticity and tidal loading. In the equilibrium theory of tides for the Moon (and Sun), the attraction on the surface is the difference between the attraction of the Moon at that point and the Moon's attraction at the center of the Earth. The tidal effect of the Moon is approximately three times the tidal effect of the Sun. Both are computed using the same equations. Gravity is measured along the normal to the Earth's surface. The difference between the Moon's attraction at the center of the Earth and the Moon's attraction at the surface in the direction of the normal is expressed as

$$g_m = \frac{GM_m}{d^2} \cos\theta' - \frac{GM_m}{R^2} \cos\theta, \qquad (2.11)$$

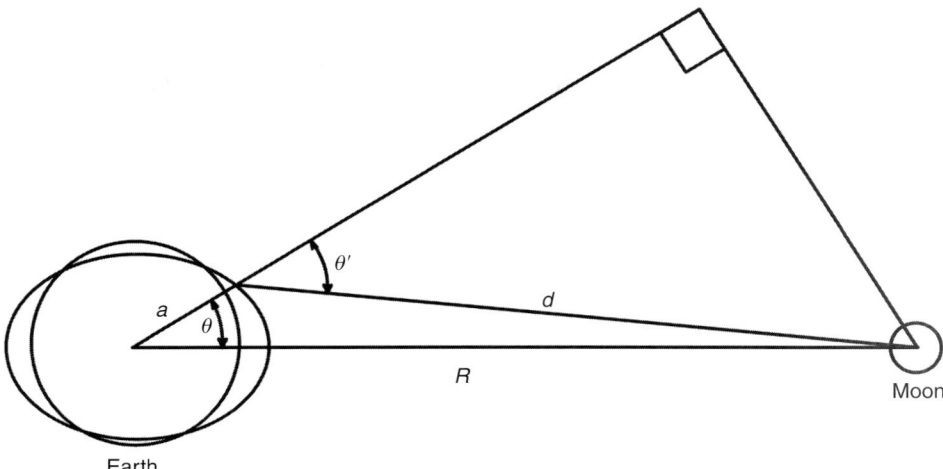

Earth

Figure 2.5 Description of geometry used in deriving the Moon's tidal attraction.

where the terms are given in Figure 2.5. The projection of R onto the direction of attraction, which is an extension of the radius a, is equal to a plus the projection of d. This relation can be used to eliminate the angle θ' and give

$$g_z = \frac{GM_m}{R^2}\left[-\cos\theta + \frac{R^3\cos\theta - aR^2}{d^3}\right].\tag{2.12}$$

Next, the law of cosines is used to change d to a form suitable for polynomial expansion in terms of powers of a/R,

$$g_m = \frac{GM_m}{R^2}\left[-\cos\theta + \frac{1}{R^3}\left(R^3\cos\theta - aR^2\right)\left(1 - \frac{3}{2}\left(\left(\frac{a^2}{R^2} - \frac{2aR}{R^2}\cos\theta\right) + \cdots\right)\right)\right].\tag{2.13}$$

After evaluating the first terms of the expansion, the first cosine term cancels out and the ratio a/R can be factored out of the remaining expression. Then, using the identity, $2\cos^2\theta = \cos 2\theta + 1$, the first approximation to the attraction of the Moon becomes

$$g_m = \frac{GM_m a}{R^3}\left(\frac{1}{2} + \frac{3}{2}\cos 2\theta\right).\tag{2.14}$$

The cosine function with twice the angle shows that the tides peak on the far and near sides of the Earth. For the rotating Earth, the relation gives two high tides per day over most of the surface.

The magnitude of the Earth tides can be estimated from the extreme values of $\cos 2\theta$ and expressing the lunar tide in terms of the ratios of the mass of the Earth and Moon. The expression becomes

$$g_m = \frac{3GM_m}{R^3} = 3\frac{GM_eM_ma^3}{a^2M_eR^3} = 3g_z\frac{M_m}{M_e}\frac{a^3}{R^3}, \tag{2.15}$$

where the g_z is the gravitational attraction of the Earth. The ratio of the mass of the Moon to the mass of the Earth is about 0.012 and the ratio of the distances is about (1/60). Using a value of 10 m/s^2 for the Earth's attraction, the lunar tidal attraction is on the order of 0.16 mGal. The corresponding value for the Sun is 46 percent of the lunar attraction, or 0.07 mGal. The combined effect, which could only be realized when the Sun and Moon are in alignment would be about 0.23 mGal. With the added tidal deformation of the surface, the additional factor of 1.2 would increase this range to 0.28. The exact value of the factor of 1.2 to compensate for elastic deformation is not critical in most gravity data reductions because application of the instrument drift correction will remove most residual variations. In computation, the relative position of the Moon and Sun are computed from astronomical equations and the time noted for the reading.

After the gravity tides are removed, the readings are corrected for meter drift and other smoothly varying changes in the gravity field. At this point the drift includes all variations in the reading over the duration of the survey. Drift in a meter is found by repeating readings periodically at reference points, typically a base station. The base station may be tied into a regional network of base stations if it is necessary to reference the data to a common datum such as IGSN71. The National Geospatial-Intelligence Agency, Office of GEOINT Sciences (2012) maintains current information on base stations. The same reference station or pairs of different reference stations can be used for a single survey. The objective is to document the change in the meter reading with time at a scale consistent with the desired accuracy of the survey. The reading sequence can be a complex sequence of repeated values for greater accuracy or simply a reoccupation of the base station at the beginning and end of the survey for lesser accuracy. The detail of the base station reoccupation sequence depends on the stability of the meter, and the precision needed in determining the anomalies. In most modern instruments the drift is usually linear and the non-predictable temporal variations are small. Therefore reoccupying the reference or base station twice a day is typically sufficient.

Multiple base ties can be used to suppress the errors associated with individual base ties. For a well-behaved meter with relatively linear drift, the observed reading, Δg_r, can be expressed as the sum of the true reading $\Delta g_r'$ and a simple time dependent expression, such as a polynomial power series

$$\Delta g_r = \Delta g_r' + a_1 t_r + a_2 t_r^2 + \cdots. \tag{2.16}$$

For purely linear extrapolation, the power series is truncated at a_1. The number of unique gravity values, s, would be the number of readings, r, minus the number of repeated (base) stations. The equations lead to a sparse matrix in which only the base ties carry significant information. The base ties can be utilized separately to determine the coefficients, a_i, using a least squares solution, or included in a larger matrix to solve for all drift corrected values and the coefficients of drift simultaneously. Equation (2.16) for any multiple sets of base ties can be written in matrix form similar to

$$
\begin{vmatrix}
\Delta g_1 \\
\Delta g_2 \\
\Delta g_3 \\
\vdots \\
\Delta g_{r-1} \\
\Delta g_r
\end{vmatrix}
=
\begin{vmatrix}
t_1 & t_1^2 & 1 & 0 & \cdots & 0 \\
t_2 & t_2^2 & 0 & 1 & \cdots & 0 \\
t_3 & t_3^2 & 0 & 0 & \cdots & 0 \\
\vdots & \vdots & \vdots & \vdots & \vdots & \vdots \\
t_1 & t_{r-1}^2 & 0 & 0 & \cdots & 1 \\
t_r & t_r^2 & 1 & 0 & \cdots & 0
\end{vmatrix}
\begin{vmatrix}
a_1 \\
a_2 \\
\Delta g_1' \\
\vdots \\
\Delta g_{s-1}' \\
\Delta g_s'
\end{vmatrix}.
\tag{2.17}
$$

In Eq. (2.17) the readings at station Δg_1 and Δg_r are at the same location and would give the single value $\Delta g_1'$. More complex base station loops would be set up the same way and would be required to solve for more than one coefficient in the power series expansion of Eq. (2.16). A least squares solution of the matrix relation is not necessarily robust because a single base tie that contains a large error will adversely affect the precision of all the data. In order to minimize the contamination of the reduction by outliers the best procedure is to independently verify the fit of the drift to individual base ties and to use the matrix inversion only after outliers have been removed and the deviations from a linear drift are confirmed to be within a reasonable error range. For modern meters, the drift is typically slight and predictable with a linear fit. For older meters or meters with higher temperature sensitivity, higher-order correction terms may be appropriate. If tidal variations are included in the drift, then for time periods greater than 6 h, the second-order term should be used. For meters that are temperature sensitive, abnormal drift can be introduced under conditions of low battery power, in environments with ambient temperatures that vary widely, or in environments with temperatures close to the operating temperature of the meter.

In addition to the base station ties, data should be reacquired at a selection of field stations throughout the survey in order to obtain a level of precision for the survey. Optimally, data at 10–15 percent of the field stations scattered throughout the survey should be reacquired at random times and from different base station loops. The repeatability of these measurements documents the instrument and site-specific uncertainty in the readings. A high-precision microgravity survey should have a repeatability of ± 0.005 mGal, while a regional survey may only require a repeatability of ± 0.050 mGal.

2.6 Variation with latitude

Latitude corrections are based on the International Gravity Formula. The values of the constants in the International Gravity Formula have been changed over time in response to improvements in the accuracy and precision of the measurements of the Earth. The theoretical gravity of an ellipsoidal Earth can be expressed in closed form as

$$
\gamma = \frac{a\gamma_a \cos\phi + b\gamma_b \sin\phi}{\sqrt{(a\sin\phi)^2 + (b\cos\phi)^2}},
\tag{2.18}
$$

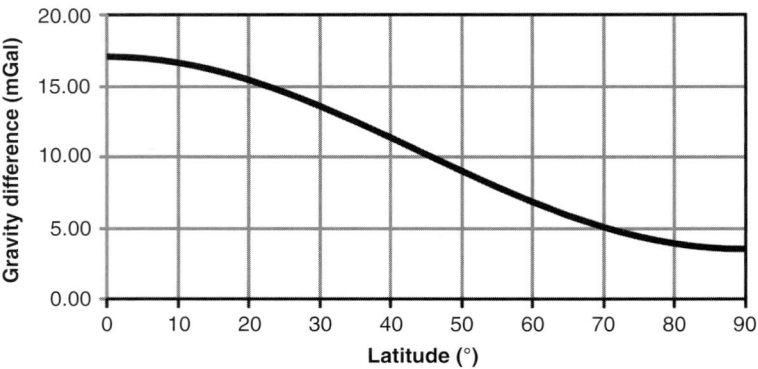

Figure 2.6 Theoretical gravity as a function of latitude.

where γ_a and γ_b are the values of gravity at the equator and pole, respectively, ϕ is the latitude, a is the radius at the equator, and b is the radius at the pole. The early forms of the Gravity Formula were expressed as a truncated series approximation in order to make computation accurate with calculators with limited significant digits,

$$\gamma = \gamma_a \left(1.0 + \alpha \left(\sin \phi\right)^2 + \beta \left(\sin 2\phi\right)^2 + \cdots\right), \tag{2.19}$$

where the coefficients α and β depend only on the mass, ellipticity, rate of rotation, and radius of the Earth.

The 1930 and 1967 International Gravity Formula are the most common reference in existing gravity anomaly maps. The 1930 International Gravity Formula corresponds to the first internationally accepted reference ellipsoid and should be used with base station values that are tied to the 1930 reference values that are determined by comparison with an absolute gravity measurement at Potsdam at that time. The 1967 International Gravity Formula is used with base stations referenced to the revised Geodetic Reference System of 1967. The coefficients for Eq. (2.19) for the two systems are

| 1930 | $\gamma_a = 9.78049$ | $\alpha = 0.0052884$ | $\beta = -0.0000059$ |
| 1967 | $\gamma_a = 9.78031846$ | $\alpha = 0.0053024$ | $\beta = -0.0000058$. |

The difference in the two equations is not simply the 14 mGal correction to absolute gravity measured at Potsdam but also incorporated a significant variation with latitude (Figure 2.6). The 14 mGal shift can cause obvious errors when, for example, used with base stations tied to the other reference system. The errors caused by comparing data reduced by the two geodetic systems are not as obvious. The difference is a function of latitude. The difference shown here varies from about 17 mGal at the equator to 4 mGal at the pole. Gravity anomalies computed using the two formulas will be the same only at the latitude where the difference is equal to the 14 mGal correction.

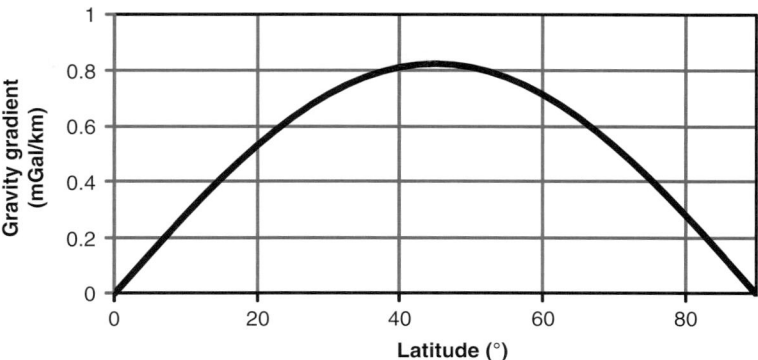

Figure 2.7 Variation of theoretical gravity gradient with latitude.

A closed form for the International Gravity Formula was recommended for the Geodetic Reference system 1980. The reference field, the World Geodetic system 1984, is given by the truncated expansion of Eq. (2.18),

$$\gamma = 9.789\,326\,7714\,\frac{1.0 + 0.001\,931\,851\,386\,39\,(\sin\phi)^2}{\sqrt{1.0 - 0.006\,694\,379\,990\,13\,(\sin\phi)^2}}. \tag{2.20}$$

Updated ellipsoids have been developed since 1980. However, they differ only slightly, on the order of a few thousandths of one mGal. The latest ellipsoids recommended by the International Union of Geodesy and Geophysics are based on the 1980 Geodetic Reference System (Moritz, 1980). Hinze et al., (2005) gives the accepted equation for the North American datum as

$$\gamma = 9.780\,326\,7715\,\frac{1.0 + 0.001\,931\,851\,353\,(\sin\phi)^2}{\sqrt{1.0 - 0.006\,694\,380\,0229\,(\sin\phi)^2}}. \tag{2.21}$$

The value of theoretical gravity, γ, expressed by any of the above equations is theoretical or normal gravity. The difference between the 1967 and 1980 geodetic reference systems is also latitude dependent and less than 0.1 mGal.

For local surveys, that is those not tied to the international geodetic reference system, the change in theoretical gravity with latitude may be removed by computing the rate of change with respect to latitude and removing the change in theoretical gravity relative to one station. The magnitude of the gradient can be computed to first order from the tangential derivative of the International Gravity Formula,

$$\frac{\partial\gamma}{r_e\partial\phi} \cong \frac{2\alpha\gamma\cos\phi\sin\phi}{r_e}. \tag{2.22}$$

The maximum is at 45 degrees latitude (Figure 2.7) and amounts to 0.8 mGal/km, or 0.01 mGal/40 ft.

2.7 Atmospheric correction

The ellipsoidal theoretical gravity given by the 1980 International Gravity Formula includes the attraction of the atmosphere so that it may also be used to define gravitational attraction in space. When the mass of the atmosphere exists in uniform elliptical shells, the atmosphere above the observation point is not reflected in the reading. Hence, a value of decreasing value with elevation must be added to the observed gravity. Wenzel (1985) presents an equation good to 0.01 mGal up to 10 km in elevation as

$$\Delta \gamma_{atm} = 0.874 - 9.9 \times 10^{-5} h + 3.56 \times 10^{-9} h^2, \tag{2.23}$$

where the height, h, is in meters above sea level and the atmospheric correction is in mGal. This is a theoretical value based on the attraction of the atmosphere as elliptical shells. The actual gradient would vary slightly, but perhaps insignificantly, from this under the influence of changes in humidity and temperature.

2.8 Free air correction

The strength of the gravity field, or gravity, decreases with increased elevation. A first-order approximation of the vertical gradient of gravity at the Earth's surface may be computed from the first term in the spherical harmonic expansion for the Earth's gravitational field. The derivative with respect to elevation takes the form

$$\frac{\partial g}{\partial h} \approx \frac{\partial}{\partial h} \left(\frac{GM_e}{(r_e + h)^2} \right) = -2G \frac{M_e}{(r_e + h)^3} \approx \frac{-2g}{r_e}. \tag{2.24}$$

The numerical value for the gravity gradient can be obtained by substituting approximate values for the equatorial radius and gravity, 6378137 m and 978032.7 mGal, respectively. Corrections for the elliptical shape of the Earth and the change in gravity from the equator to the pole increase this estimate by 0.0016 mGal/m to an average value for the vertical gradient of 0.3085 ± 0.0003 mGal/m (2.09403 mGal/ft). The small variation in the vertical gradient with latitude can be attributed to the approximately equal influence of the decrease in gradient related to decreased curvature and increase in gradient from the increase in gravity from the equator to the pole. This estimate for the vertical gradient does not include the contributions of local anomalies to the vertical gradient; however, these are small and can often be ignored in most areas of slight topography.

To a first order, the change in the vertical gradient with increased elevation can be approximated by expanding Eq. (2.24) to higher order of precision in elevation,

$$\frac{\partial g(h)}{\partial h} \approx -\frac{2g(h=0)}{r_{h=0} \left(1 + \dfrac{h}{r_{h=0}} \right)^3} = -\frac{2g(h=0)}{r_{h=0}} \left(1.0 - \frac{3h}{r_{h=0}} + \cdots \right). \tag{2.25}$$

At an elevation of 3 km the gradient is decreased from 0.3085 to 0.3081 mGal/m. The gradients in the gravity field given here are obtained from the expression for theoretical gravity. These variations in the vertical gravity gradient are small in comparison to the uncertainty in the Bouguer plate correction discussed below and changes in the curvature of the gravity field associated with anomalous density structure. Hence, a value of 0.3086 mGal/m at 45 degrees latitude is sufficient for most gravity data reductions in areas of low topography.

The magnitude of the change in gravity for a unit vertical distance is as much as 3000 times greater than the change in gravity with a comparable distance in latitude. Also, the density contrast for many typical structures in the Earth's crust have gravity anomalies on the order of 1 to 10 mGal, values comparable to the change in gravity expected for elevation differences of 3 m to 30 m. Hence, the dependence of gravity on elevation will dominate the reduction of gravity measurements and will have to be removed to observe anomalies caused by changes in the density of various structures. The free air gravity anomaly is defined as the observed value with the theoretical gravity on the ellipsoid subtracted and the data corrected to a constant elevation (usually the geoid). The equation for the free air gravity anomaly is

$$\Delta g_{FA} = g_{ob} - \gamma + \frac{\partial g}{\partial h} \Delta h. \tag{2.26}$$

The first term is the observed gravity at the observation station with time-varying components removed and its value tied to a network of absolute gravity values; the second term is theoretical gravity on the ellipsoid. The third term corrects the gravity value to an equipotential (or level) surface, usually taken as the geoid. Above sea level, the change in elevation is negative, so the third term corrects for the decrease in gravitational attraction with increased elevation. The reduction will accurately represent the gravity on that level surface only in conditions where no mass lies between the point of gravity observation and the level surface; a condition that rarely exists. Instead, the elevation correction in the free air reduction is a first approximation to downward continuation of the gravity field to the geoid. Also, the elevation is the map elevation obtained from leveling, the elevation that is posted on conventional topographic maps. The resulting gravity values on the level surface are simply those values that give the observed gravity when continued back to the elevation of observation. The free air anomaly as defined in Eq. (2.26) then represents the difference between the gravity field continued down to the geoid, which is equivalent to mean sea level, and the theoretical attraction of gravity on the ellipsoid. Theoretical gravity is referenced to the theoretical or "normal" potential field, the ellipsoid, for the Earth and the position of reference for theoretical gravity may differ from the geoid. The separation of the ellipsoid and geoid is N, the geoid height (Figure 2.8). The geoid height, N, is commonly referred to as the shape of the Earth. Although N does not include the ellipticity, it is the most relevant measure of anomalous structures and tectonic activity.

Satellite-based GPS measurements of elevation have largely replaced map and surveyed elevations for regional surveys. These elevations are referenced to the ellipsoid and may differ from surveyed or map elevations. The precision in elevation required for small-scale and microgravity surveys still requires leveling techniques so that in these surveys the

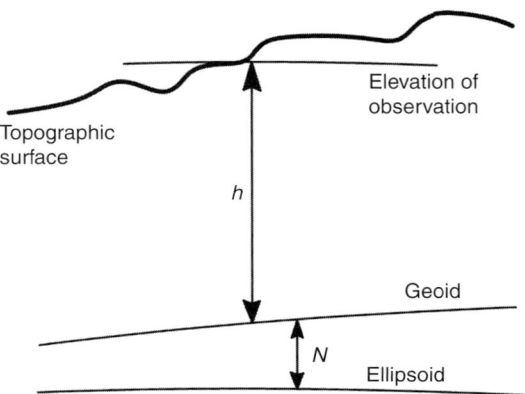

Topographic
surface

Elevation of
observation

h

Geoid

N

Ellipsoid

Figure 2.8   Relation between the geoid and ellipsoid.

difference between the position of the geoid and ellipsoid is not relevant. However, because
many of these measurements may eventually be incorporated in global data sets, (Hinze
et al., 2005) recommend a revision to the free air correction that uses a second-order
approximation for the change in theoretical gravity based on the GRS80 ellipsoid with
heights determined relative to the ellipsoid. The equation for the vertical gradient is taken
from Heiskanen and Moritz (1967, equation 2–124), which is a power series expansion to
second order expressed in terms of the principal radii of curvature of the ellipsoid. It is
good for small elevations above the ellipsoid, which would include all terrestrial gravity
measurements. The ellipsoidal elevation correction equation is

$$\delta\gamma = -\frac{2\gamma_a}{a}\left[1 + f + m + \left(-3f + \frac{5}{2}m\right)\sin^2\phi\right]h + \frac{3\gamma_a}{a^2}h^2, \qquad (2.27)$$

where a is the semi-major axis, γ_a is gravity on the semi-major axis, f is the flattening, and
h is the elevation. The constant m is $\omega^2 a^2 b / GM$ where ω is the angular velocity, b is the
semi-minor axis and GM is the geocentric gravitational constant. For the 1980 ellipsoid,
the evaluation of the constants is given in Hinze et al. (2005) as

$$\delta\gamma(h) = -(0.308\,7691 - 0.000\,4398\sin^2\phi)\,h + 7.2125 \times 10^{-8}\,h^2, \qquad (2.28)$$

where the ellipsoidal height is in meters and the gravity correction is in mGal. The gradient
depends slightly on latitude, but the second term is greater than the latitude effect at high
elevations. At 10 km in elevation the difference between the first-order Eq. (2.25) and
Eq. (2.28) is on the order of 10 mGal.

When using ellipsoidal (i.e. GPS) elevations and corrections according to Eq. (2.28),
Eq. (2.26) should be written as

$$\delta g_{FA} = g_{ob} - \gamma + \delta\gamma(h), \qquad (2.29)$$

where the equation computes what traditionally has been called the gravity disturbance,
δg_{FA}. Heinz et al. (2005) suggests calling this the ellipsoidal free air anomaly, which with
more frequent use will reduce in time to simply the free air anomaly.

2.9 The simple Bouguer correction

While the free air correction removes the theoretical variation of gravity with respect to elevation in free space, the resulting free air anomalies are strongly correlated with the local topography. The mass between the station and the ellipsoid and the masses within mountains and valleys turn out to be strong contributors to the gravitational attraction at a point. The density contrast of a mountain or valley is approximately the same as the density of the rock in the mountain and, for that reason, variations in the thickness of mass below a point have a much greater influence than the smaller density contrasts of geologic structure.

In order to investigate the perturbations in the gravity field associated with anomalous density structure, the gravity anomalies should only represent anomalous densities associated with different rock units. In efforts to model geologic structures, variations in attraction related to the topography should be avoided. The objective of the Bouguer plate correction is to remove the influence of the mass between the station and the ellipsoid. Sea level or the geoid is the datum to which most older gravity data have been reduced. When only a flat plate is used to account for this mass, the anomaly is the simple Bouguer anomaly. Where the topography deviates from a flat plate, the influence of the mass in these topographic features is removed as a topographic correction. When the effects of both the elevation, in terms of a flat plate, and the influence of mountains and valleys are removed, the resulting anomaly is the complete Bouguer anomaly.

A pie-shaped section of a vertical cylinder as shown in Figure 2.9 may be used to derive useful equations for the attraction of many shapes, including a flat plate and irregular topography. The flat plate is used in the Bouguer plate correction and the segment of a vertical cylinder can be used in the topographic correction. The vertical component of an infinitesimal element of attracting mass in a vertical cylinder can be expressed in cylindrical coordinates (s, α, z) and integrated over the total volume. In Figure 2.9, the integration is over a pie-shaped portion of thickness b, radius a, and contained between angles α_1 and α_2. The integral takes the form

$$\Delta g_z = G \int\limits_{\alpha_1}^{\alpha_2} \int\limits_0^a \int\limits_0^b \frac{\rho(s, \alpha, z)s\,(c - z)\,dz\,ds\,d\alpha}{\left(s^2 + (c - z)^2\right)^{3/2}}, \tag{2.30}$$

where the denominator is the distance from the infinitesimal element of integration to the point at c on the z-axis. In order to simplify the solution, the density will be assumed to be constant over the volume of the cylinder and moved outside the integration. The integration about α is evaluated first as

$$\Delta g_z = G\rho(\alpha_2 - \alpha_1) \int\limits_0^b \int\limits_0^a \frac{(c - z)\,s\,ds\,dz}{\left(s^2 + (c - z)^2\right)^{3/2}}. \tag{2.31}$$

For a full cylinder, the difference in the angles of the sides would be 2π. Successive integration with respect to s and z, and evaluation at the limits give the gravitational

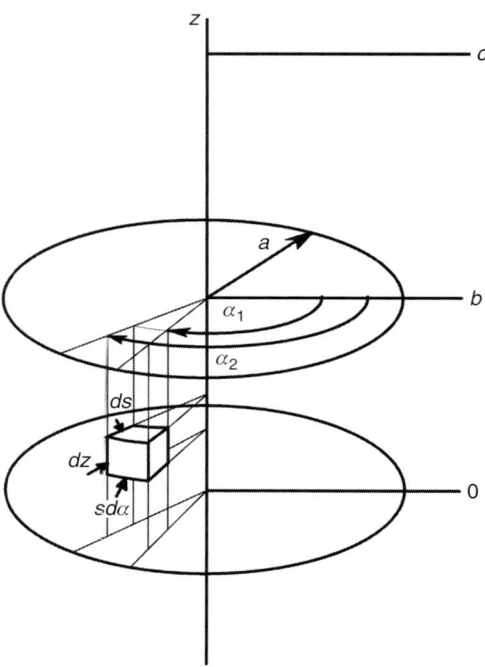

Figure 2.9 Geometry for finding the attraction of a vertical disk.

attraction in the vertical direction of a pie-shaped slice of a vertical cylinder on its axis. The equation is

$$\Delta g_z = (\alpha_2 - \alpha_1)G\rho \left(b + \sqrt{a^2 + (b - c)^2} - \sqrt{a^2 + c^2} \right). \qquad (2.32)$$

If the observation point is at the top surface of the cylinder at $z = b$ instead of $z = c$ then the equation simplifies to

$$\Delta g_z = (\alpha_2 - \alpha_1)G\rho \left(b + a - \sqrt{a^2 + b^2} \right). \qquad (2.33)$$

This is the gravitational attraction on the axis of a sector of a cylinder. To get the attraction of a flat plate, the angle is expanded to 2π and the limit is evaluated as a goes to infinity. The result is

$$\Delta g_{Bp} = 2\pi G\rho b, \qquad (2.34)$$

where the subscript Bp denotes that the equation is the Bouguer plate correction for an elevation b. Substituting for the values of G and a conventional crustal density of 2.67 g/cm^3, the correction is $\Delta g_{Bp} = 0.1119b$ mGal, where b is in meters. The rocks near the surface do not always conform to the standard density of 2.67 g/cm^3, which is based on the assumption that an appropriate reduction density would be the same as the density of granitic rocks. Deviations of the reduction density are examined in Chapter 6. For comparisons of regional gravity maps, the standard reduction density of 2.67 g/cm^3 should be used. However, for interpretation of near-surface structures, a density consistent with the density of the shallow rocks should be used.

If leveling data are used for elevation control, the gravity data are reduced down to the geoid and b is the same as Δh. If GPS elevations are used, the term ellipsoidal Bouguer anomaly would be appropriate as the height, $b = \Delta h$, would be the distance between the observation station and the ellipsoid. The attraction of mass below the station is subtracted from the free air reduction to obtain the simple Bouguer anomaly,

$$\Delta g_B = g_0 - \gamma + \frac{\partial g}{\partial h}\Delta h - \Delta g_{Bp}. \tag{2.35}$$

In many reductions the elevation and Bouguer plate corrections are combined to give a simpler relation for standard density as

$$\Delta g_B = g_0 + 0.196\Delta h, \tag{2.36}$$

where h is the elevation in meters.

A revised Bouguer correction has been proposed (Hinze et al, 2005) to correct for the curvature of the Earth. The height of the Bouguer slab, b, in Eq. (2.34) is replaced by the closed-form formula for a spherical cap of radius 166.7 km. The magnitude of this revised correction differs by ± 1.5 mGal from the traditional Bouguer slab correction, depending on the elevation of the measurement. The appropriate equation from LaFehr (1991) for the additional Bullard B correction for curvature is

$$\Delta g_{Bp} = 2\pi G\rho(\mu h - \lambda R), \tag{2.37}$$

where μ and λ are dimensionless coefficients defined by LaFehr (1991), R is the radius of the Earth, $R = R_0 + h$, at the station and R_0 is the mean radius of the Earth. Values for Eq. (2.37) are given in LaFehr (1991, Table 1) and range from zero at zero elevation to a maximum at 2075 m of 1.518 mGal and then falling off to -1.5 mGal at a height of 5000 m. The values of the Bullard B correction given in LaFehr (1991, Table 1) can be computed to within 0.001 mGal using the relation

$$\Delta g_{Bp} = -0.352\,7404h^2 + 1.463\,65h, \tag{2.38}$$

where h is in kilometers in the range of 0 to 4 km of elevation.

2.10 Terrain corrections

The equation for the attraction of a segment of a vertical cylinder has traditionally been used to compute the effects of terrain on the gravitational attraction. The reference elevation is the elevation of the observation point because the terrain correction compensates for the topography that deviates from the Bouguer plate correction at that point. The area surrounding the gravity observation point is divided into segments that typically increase in size with distance. The Hammer template (Hammer, 1939) has segments divided so that each segment contributes equally to the terrain correction for each meter of elevation

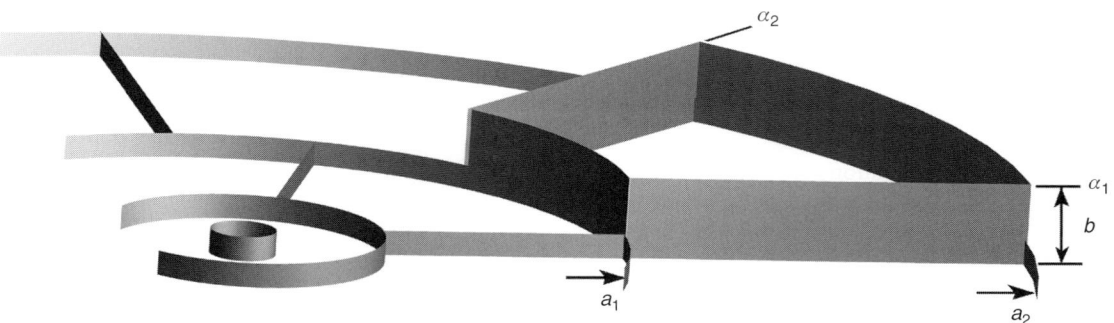

Figure 2.10 Circular pattern used in the Hammer topography correction.

above (or below) the observation station. An increase in compartment size with distance is required to maintain a constant contribution to the terrain correction for each meter of elevation difference between the elevation at the observation point and the average elevation in the compartment. If the observation point is at the bottom surface of the cylinder, at $z = 0$ instead of $z = c$ in Eq. (2.32) and Figure 2.10, and the attraction for a radial distance of a_1 is subtracted from the attraction for a radial distance of a_2, the attraction for the compartment becomes

$$\Delta g_z = (\alpha_2 - \alpha_1) \, G\rho \big(a_1 - a_2 + \sqrt{a_2^2 + b^2} - \sqrt{a_1^2 + b^2}\big). \qquad (2.39)$$

This is the gravitational attraction on the axis at the base of a section of a cylinder. Historically, a template was drawn as defined in Hammer (1939) on a transparent overlay at the scale of a map with elevation contours. Then, with the template centered on the observation point, the difference between the elevation at the observation point and the estimated average elevation in the compartment is found and used to compute the terrain contribution from that compartment using Eq. (2.39). The contribution of the other compartments are computed and summed to give the total terrain correction. The size of the compartment may be adjusted to fit any topographic feature. The computation is repeated for each observation point. If this process is performed by hand, it is tedious and not advised where digital data are available. The template does illustrate that small topographic irregularities near the reading point can have consequences comparable to significantly greater topography at distance.

The topographic correction is a strong function of distance. Consequently, near-by topographic features have the greatest influence. A road cut or vertical escarpment can be approximated by a half-cylinder obtained by setting the difference in angles to π radians in Eq. (2.33). A reading at the base of a 5 m high vertical road cut will have a topographic correction of 0.28 mGal. However, if one moves 30 m away from the road cut, the topographic correction will be less than 0.03 mGal. For gravity data with a desired precision of 0.1 mGal, this shows the importance of choosing reading sites with minimal topography in the immediate vicinity.

The terrain correction is always positive (Figure 2.11). For a mountain, the force of the mountain's mass above the observation point is not included in the Bouguer plate correction.

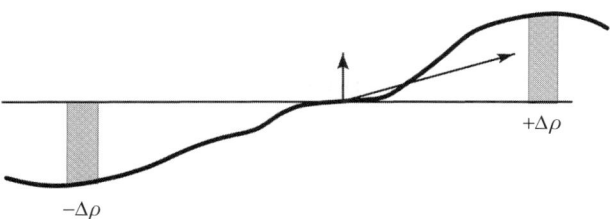

Figure 2.11 Terrain correction is always positive.

This excess mass has a vertical component at the observation point. Figure 2.11 shows a small section of a mountain. The attraction of each section of the mountain is directed toward each section from the observation point. The component of the mountain's attraction in the vertical direction is upward, decreasing the value of gravity at the observation point. This reduces the measured gravity and hence requires a positive terrain correction. In a valley, the lack of mass represents a negative deviation in density from the Bouguer plate reduction. The terrain correction for a valley compensates for the lack of attraction that was originally assumed in the Bouguer plate reduction. The valley, in effect, does not exert a downward force and it also reduces the measured attraction at the observation point. The required terrain correction is again positive. When the terrain correction is added to the simple Bouguer anomaly, the resulting anomaly is the complete Bouguer anomaly and is defined by

$$\Delta g_B = g_0 - \gamma + \frac{\partial g}{\partial h} h - \Delta g_{Bp} + \Delta g_{\text{terr}}. \tag{2.40}$$

For digital elevation data Cogbill (1990), the elevation values represent areas that have sides that are small in comparison to the distance to the observation point. With this restriction, the computation of the topographic correction may be approximated by the attraction of a vertical column with its mass concentrated on the axis. Only the attraction of points that are adjacent to the observation point and that have significant topography would need to be computed using a more exact representation for the topography. We start with the equation for the attraction of an element of a rectangular column,

$$\Delta g_z = G \Delta \rho \int_{\frac{-\Delta x}{2}}^{\frac{\Delta x}{2}} \int_{\frac{-\Delta y}{2}}^{\frac{\Delta y}{2}} \int_0^h \frac{d\xi \, d\eta \, d\varsigma}{\left(r^2 + \varsigma^2\right)^{3/2}}, \tag{2.41}$$

where r is the distance from the observation point to the rectangular column. The first two integrations are over the area of the top of the column. Because r is large compared to the sides of the column, the mean value theorem can be used and the integral over ς gives

$$\Delta g_z = -G \Delta \rho \Delta x \Delta y \left(\frac{1}{r} - \frac{1}{\left(r^2 + \varsigma^2\right)^{1/2}} \right), \tag{2.42}$$

where r is now the distance to the center and the sides of the rectangular area are Δx and Δy. The height z is the difference between the elevation of the observation point and the mean

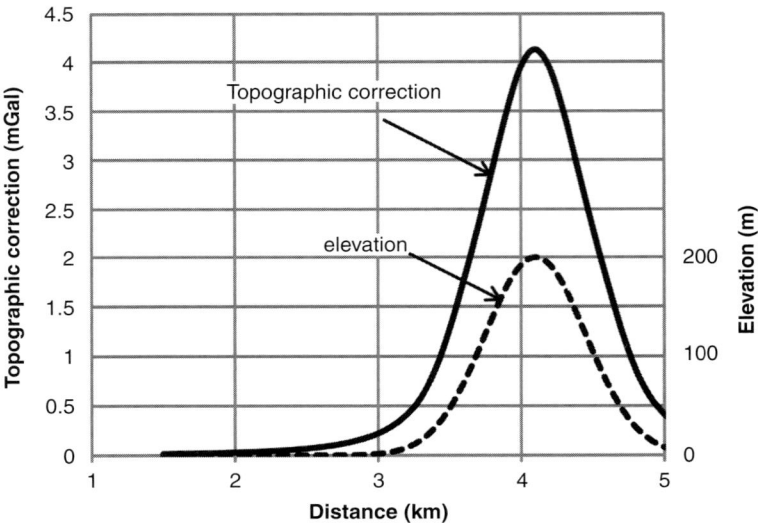

Figure 2.12 Comparison of the topographic correction (solid line) for a simulated elevation (dashed line).

elevation in the rectangular area. The terrain correction is the sum of this attraction over all elevation points. As an example, Stone Mountain, Georgia, is an isolated exposed granite dome with a peak elevation of 200 m above the surrounding area. A gravity reading at the center of the top of Stone Mountain will have a topographic correction of approximately 4.2 mGal. The topographic correction at the base of Stone Mountain (Figure 2.12) is around half the peak value and the magnitude of the terrain correction decreases with distance from the mountain. In mountains with topographic relief of 1000 to 2000 m the terrain correction can be on the order of 10 mGal.

Traditionally, regional gravity surveys have had minimal (if any) terrain corrections applied, mainly due to the extensive processing time and lack of digital elevation models at the time of the survey. Modern computers and LiDAR-based elevation models have allowed for more efficient and accurate terrain corrections to be applied to datasets. However, the procedures for applying terrain corrections vary greatly, and one should use caution when comparing datasets with different terrain correction parameters. Hinze et al. (2005) suggest a standardized, three-phased approach to terrain corrections that uses near-station user-obtained elevations to a distance of 100 m from the station, high-resolution digital elevation models to a distance of 895 m from the station, and more coarsely sampled terrain models to a distance of 166.7 km from the station.

2.11 Isostatic anomaly

At the observation station the complete Bouguer anomaly removes the attraction of the mass above the reduction elevation, usually mean sea level. However, if Bouguer anomalies

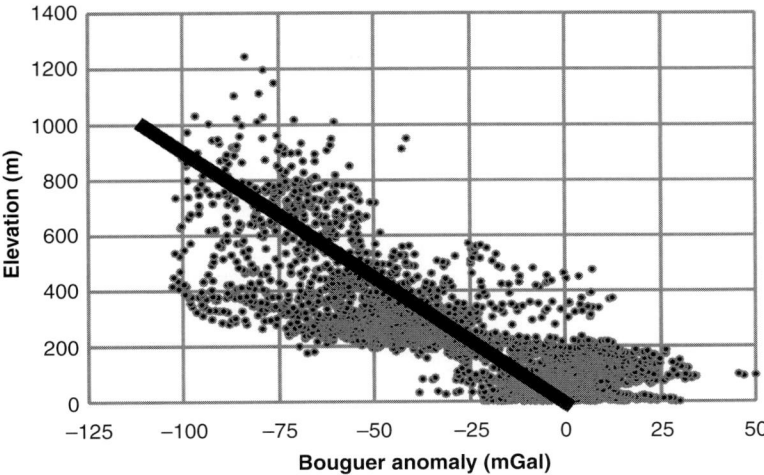

Figure 2.13   Observed variation of Bouguer anomaly with increasing elevation in the Southern Appalachians. The heavy line is the magnitude of the Bouguer plate correction as a function of elevation.

from a large area covering observations both at sea level and at elevation in mountainous regions are plotted against the elevation of the observation point, the Bouguer anomalies become more negative with increasing elevation. This relation can be observed for gravity data covering the southeastern United States, which includes the Southern Appalachian Mountains with elevations approaching 2000 m and the Coastal Plain, which is near sea level. The Bouguer anomalies decrease at a rate that is equivalent to the mass removed in the Bouguer reduction. Terrain corrections that are on the order of 5 to 10 mGal at the higher elevations would not significantly change this general relation because they would fall within the one standard deviation of the data plotted in Figure 2.13. Also, world-wide gravity measurements confirm that the Bouguer anomalies are positive in proportion to the depth over oceans as well as negative in proportion to the height of the mountains. The correlation of elevation with the variations in the Bouguer anomalies on a large scale must be caused by variations in the density of the Earth's lower crust and upper mantle because the Bouguer plate correction has removed the effects of mass between the observation point and sea level. These observations (Figure 2.13) indicate that there is an anomalous increase in low-density material under mountains and an increase in high-density material under the oceans. Figure 2.13 also demonstrates that the magnitude of the negative Bouguer anomaly is on average the same as the Bouguer plate correction. The implication of this relation is that the anomalous mass below sea level is on average equal to the mass above sea level. In agreement with the principle of isostatic equilibrium, the deficiency in mass below a mountain is on average equal to the weight of the mountain above sea level. The weight of a topographic feature above sea level or the decreased weight of sea water relative to continental crust is compensated at some depth by an approximately equal mass deficiency or excess.

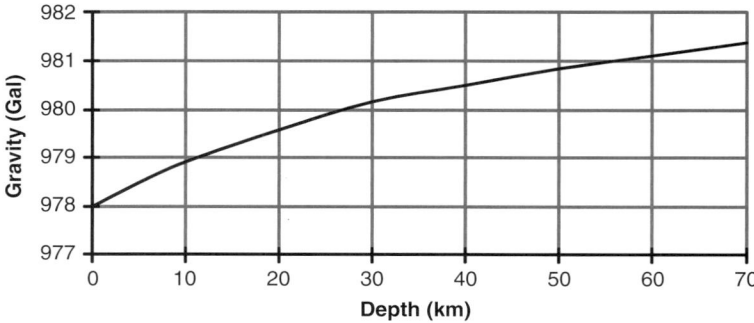

Figure 2.14   Variation of gravity with depth in the crust.

The principle of isostatic equilibrium explains the correlation of Bouguer anomalies with elevation. Isostatic equilibrium is a condition where the buoyant pressure of a mass deficiency at depth balances the topography and anomalous density structure near the surface of the Earth's crust. In isostatic equilibrium the forces tending to support a mountain balance those tending to depress the Earth's surface. Isostatic equilibrium is achieved at some depth when the pressure at that depth, the depth of compensation, is a constant independent of position. For the Earth, the depth of compensation is at or above the athenosphere, a zone of low shear-wave velocity and weakness in the upper mantle. Deviations from isostatic equilibrium indicate stresses in rigid portions of the crust and upper mantle or tectonic movements. The weight of the mass above the depth of compensation is the same for all areas in isostatic equilibrium. The pressure at the depth of compensation can be expressed by the integral

$$C = \int_{-D}^{h} \rho(z)\, g_z(z)\, dz, \tag{2.43}$$

where C is a constant, and the limits extend from the depth of compensation $-D$ to the elevation h. The density of the Earth is a function of depth and position. The depth dependence is largely a function of rock composition and pressure. Tectonic features and processes control lateral variations in density.

Unlike the gravitational attraction above the surface, which decreases at a rate of 0.3085 mGal/m, the gravitational acceleration in the Earth decreases at a rate of about 0.1 mGal/m (Figure 2.14). The gravitational attraction actually increases less than 0.35 percent in the top 60 km. When measuring below the surface such as in a tunnel or well the Bouguer reduction has to account for mass both below and above the reading point. Likewise, measuring a vertical gradient in a building must take account of the building mass both above and below the gravity meter. By way of explanation, in the shallow low-density crust the decrease in attraction with depth caused by the reduced fraction of the Earth's total mass below a given depth is less than the increase in gravitational attraction caused by the decrease in distance from the center of the Earth. For this reason a constant value for the gravitational attraction may generally be assumed for the Earth's mantle. In the near-surface

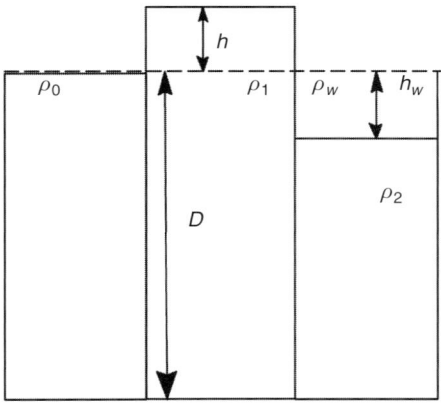

Depth of compensation

Figure 2.15 Pratt–Hayford system for computation of the isostatic correction.

and crust, the gradient is dependent on the rock density and would need to be removed to use gravity to identify anomalous density structures.

One of the great debates of the late 1800s concerned the correct model for isostatic compensation. Although, it was not isostatic anomalies that sparked the debate, but a discrepancy between computed and measured horizontal deflections of the direction of gravity. A measurable increase in the deflection of the vertical toward the Himalayan Mountains had been observed at stations close to the mountains. Calculations by J. H. Pratt for the attraction of the mass of the Himalayan Mountains overestimated the deflections by a factor of five. In order to explain the difference, a negative mass anomaly, a compensating mass, must exist near or under the mountains.

The model attributed to J. H. Pratt shown in Figure 2.15 was subsequently put into mathematical form by J. F. Hayford for geodetic computations. In the Pratt–Hayford system the mountains are assigned a lower density in proportion to their height. Arguments for the assumption of lower densities in mountains included a belief at the time that many rocks in mountains were lower in density than those found at lower elevations. In the mathematical formulation of the model, the lower density extends down to the depth of compensation. Hence, the lower density below sea level can compensate for the excess mass above sea level. Using Eq. (2.43) we integrate from the depth of compensation $-D$ to the elevation h in the three columns illustrated in Figure 2.15 to get the relation

$$\rho_0 D = \rho_1 (D + h) = \rho_2 (D - h) + \rho_w h, \tag{2.44}$$

where the terms are as given in Figure 2.15. In mountainous regions the density contrast is given by

$$\Delta \rho = \rho_0 - \rho_1 = \frac{h \rho_0}{D + h}, \tag{2.45}$$

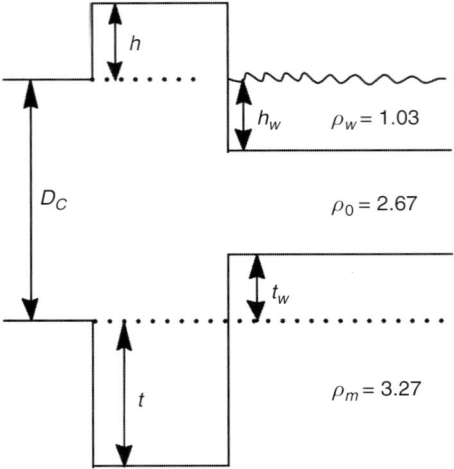

Figure 2.16 The Airy–Heiskanen system for computation of the isostatic correction.

and in the oceans the density contrast can be computed from

$$\Delta \rho = \rho_2 - \rho_0 = \frac{h_w (\rho_0 - \rho_w)}{D - h_w}. \tag{2.46}$$

The conventional assumption for standard crustal density, $\rho_0 = 2.67$ g/cm^3, and a depth of compensation will give the perturbed density under mountains. An appropriate depth of compensation for use in computation would be one that minimizes the gravity anomaly. The solution for depth of compensation over various continental regions typically gives values in the range of 100 to 200 km. However, the Pratt–Hayford system (Figure 2.15) is a simplistic mathematical formulation and compensation in the Earth is largely achieved at shallower depths. The depth of compensation typically exceeds the crustal thickness because the actual compensation is spread out over an area, not just under the topographic feature as assumed in local compensation.

The model attributed to G. B. Airy (Figure 2.16) is a more realistic representation of crustal structure. This model was given a mathematical treatment for application to geodetic studies by Heiskanen. The Earth's crust in this model is given a constant standard density of 2.67 g/cm^3 and floats on a denser upper mantle of density 3.27 g/cm^3. The higher mountains, with more mass above sea level, will protrude deeper into the mantle. The density contrast between the crust and upper mantle provides the buoyancy needed to support the mountains above sea level. Again, using Eq. (2.43) by integrating over the anomalous mass, which in the Airy–Heiskanen model is from the base of the crust to the elevation of the mountains, we obtain

$$\rho_0 g D_c = \rho_0 g h + \rho_0 g D_c + (\rho_0 - \rho_m) g t = (\rho_w - \rho_0) g h_w + \rho_0 g D_c + (\rho_m - \rho_0) g t_w \tag{2.47}$$

for the three columns in Figure 2.16. Using the assumed standard densities for the crust, mantle, and oceans, we get

$$t = \frac{\rho_0}{\Delta\rho}h = 4.45ht \tag{2.48}$$

for the areas above sea level and

$$t_w = \frac{\rho_0 - \rho_w}{\rho_m - \rho_0}h_w = 2.73h_w \tag{2.49}$$

for areas over the oceans. The Airy–Heiskanen system is recommended by Hinze et al. (2005). The recommended crustal thickness at sea level elevations is 30 km and the recommended density contrast with the mantle is 0.3 g/cm^3.

The Pratt–Hayford and Airy–Heiskanen systems are idealistic and simple. They both assume that compensation is achieved in a vertical column. This assumption is referred to as local compensation. Local compensation can be achieved only where the topography varies slowly within distances comparable to the depth of compensation or in studies that average areas of dimensions greater than the depth of compensation. The topographic features with dimensions less than the thickness of the crust can be supported by the strength of the crust, or at least that portion of the crust with sufficient strength to support the load. The topographic load will then, through elastic bending, depress a portion of the crust into the mantle, and the compensation will be distributed over an area and depth range that is dependent on the elasticity and composition of the crust and upper mantle.

The attraction of local isostatic compensation may be computed in the same way as the topographic correction. Either Eq. (2.42), the equation for a vertical line, or Eq. (2.31) for a segment of a vertical cylinder, may be modified to fit the appropriate model and elevation data. For the Airy model, the segment of a vertical cylinder between radius a_1 and a_2 would give

$$\Delta g_z = (\alpha_2 - \alpha_1)G\rho\left[\sqrt{a_2^2 + D_c^2} - \sqrt{a_1^2 + D_c^2} - \sqrt{a_2^2 + (D_c + t)^2} + \sqrt{a_1^2 + (D_c + t)^2}\right], \tag{2.50}$$

where D_c and t are as defined in Figure 2.16. As in the terrain correction, the computation of the effects of compensation must be made independently for each point. The isostatic correction is the summation of the effects of the compensation for all the surrounding topography. The isostatic anomaly is then computed from the complete Bouguer anomaly by the addition of the computation for the effects of compensation

$$\Delta g_I = g_0 - \gamma + \frac{\partial g}{\partial h}h - \Delta g_{Bp} + \Delta g_{\text{terr}} + \Delta g_{\text{isos}}. \tag{2.51}$$

A comparison of the gravity anomalies for the isostatic compensation computed by the Airy and Pratt models demonstrates that the difference is less than 4.0 mGal for a one-kilometer step in elevation. These differences compare to a typical variance in the gravity anomalies from geologic structures in the crust that are on the order of 5 to 15 mGal and a topographic correction of 5 to 10 mGal for mountains of one kilometer in elevation and a typical Bouguer anomaly of −100 mGal. One might conclude that the Airy and Pratt

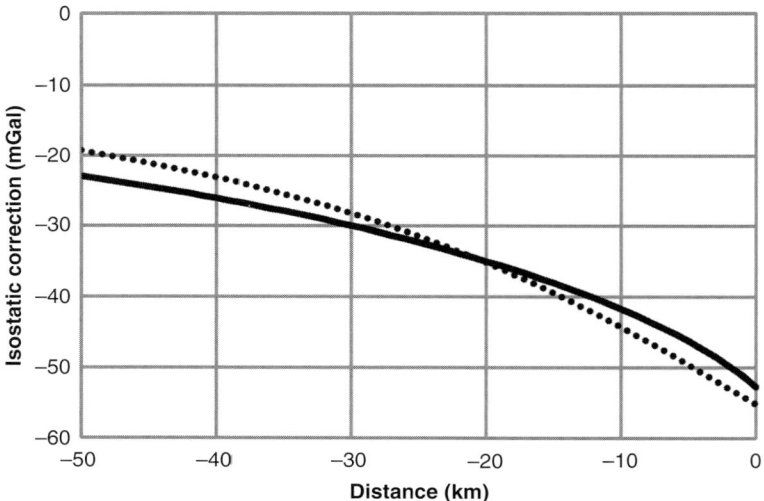

Figure 2.17 Comparison between Airy and Pratt compensation models.

concepts applied to isostatic anomaly computation give results within 5 percent. Uncertainty in the depth of compensation, which is found by minimizing the isostatic anomaly, would also give differences on the same order of magnitude as the difference between these two computational methods (Figure 2.17).

The computation of the isostatic anomalies may include corrections for deviations from a plane surface. The curvature of the Earth has a minor effect on the width of columns used for isostatic correction. Hence, the isostatic anomalies can include the influence of the convergence of the sides of each column toward the center of the Earth. However, such corrections are small compared to uncertainties in anomalies and density structure. Also, the Bouguer plate correction and other corrections based on distant measurements can be corrected for the curvature of the Earth.

2.12 Characteristics of the different reductions

The removal or shifting of mass inherent in the computation of most gravity reductions will modify the appearance and meaning of the derived anomalies. Also, the various gravity reductions have an impact on the interpretation of the shape of the Earth and the mass distribution in the crust and mantle. The appropriate gravity reduction will depend on the objective of the analysis. The shape of the Earth, as defined by the geoid, can be inferred approximately from the sea level surface, but cannot be measured directly over land. The geoid may be indirectly computed from gravity anomalies. Consequently, gravity reductions that remove or move mass can influence geoid computations. Each reduction has its own cogeoid, the difference between the true geoid and the geoid computed using gravity data computed using that reduction. The following summarizes the utility and characteristics of the most common gravity data reductions.

When using gravity values obtained using different reductions in conventional equations to define the shape of the Earth, the result is a cogeoid unique to each reduction. The cogeoid differs from the actual geoid by a distance referred to as the indirect effect. The cogeoid is defined by

$$N^c = N - \Delta N, \tag{2.52}$$

where N^c is the cogeoid, N the geoid, and $\Delta N = \Delta W / \gamma$ is the indirect effect that is related to the change in the potential, ΔW, caused by a particular reduction.

Free air. The free air anomaly corrects only for the vertical gradient of the gravity field and is the simplest to compute. Free air anomalies when averaged over an area on the order of one degree square are relatively small, but at a local scale of a few kilometers they can be rough and strongly correlated with the local topography. The roughness at local scales is caused by mass distributions associated with the local topography. At local scales free air anomalies have limited geophysical significance because the topographic signature dominates over anomalies from variations in density. At regional scales the free air anomalies may be considered an approximation to isostatic anomalies. The indirect effect for the free air reduction is relatively small and is the consequence of using gravity on the ellipsoid rather than the geoid as a reference for theoretical gravity. If ellipsoidal heights (GPS elevations) are used the reduction is to the gravity disturbance and the indirect effect is zero.

Bouguer anomaly. The Bouguer anomaly is the most useful anomaly for the interpretation of density structure in the shallow crust of the Earth. This is the anomaly best used for interpretation of geologic structures. By removing the influence of mass between the station and sea level (or ellipsoid), the Bouguer anomalies remove the direct influence of the mass below the station from the free air anomalies. The complete Bouguer anomaly also includes a correction for terrain not directly under the station. The complete Bouguer anomaly thus should correspond directly to density structure. In mountainous regions, isostatic compensation at depth causes the Bouguer anomaly to become increasingly negative with increasing elevations. The cogeoid for the Bouguer anomaly is on the order of 10 times the geoid because of the magnitude of the mass removed. Thus, the Bouguer anomalies are not appropriate for computations of the geoid.

Faye anomaly. The Faye anomaly is the free air anomaly with a terrain correction. The cogeoid for the Faye anomaly is very small and, hence, the Faye anomaly is commonly used to compute the geoid from gravity data.

Isostatic anomaly. Isostatic anomalies are Bouguer anomalies with the topographic mass moved to some other position. Isostatic anomalies are smooth, like the Bouguer anomalies and are typically small, like the free air anomalies. Isostatic anomalies require more computational detail and the location of the theoretically replaced mass may differ from the actual isostatic compensation in the Earth. The cogeoid is small and depends on the placement of the replaced mass.

Condensation reduction of Helmert. In the condensation of Helmert reduction the topography is condensed to a thin layer at the elevation of the geoid. It may be considered an isostatic reduction using the Pratt–Hayford system with a zero depth of compensation. The free air reduction is in one sense a limit of the condensation reduction because the mass above the geoid is essentially canceled out by the compensating mass at the geoid. The cogeoid is comparable to that of the free air reduction.

Inversion reduction of Rudzki. The inversion reduction of Rudzki replaces the topographic mass by an inverse topographic mass. This is the only gravimetric reduction that does not change the geoid. Hence, the cogeoid for the Inversion Reduction is zero, but the gravity anomaly outside Earth is changed. Some schemes correct for the convergence of Earth's radius.

Field acquisition of gravity data

3.1 Introduction

Gravity data acquisition is an exercise in the documentation of detail and common sense. The practicality of developing detailed survey plans beforehand depends on the type of survey. A survey over a small area where the locations of data points can be determined in advance could benefit from a site visit and detailed preplanning. However, for more widely spaced data, such as in regional surveys for crustal-scale structures, the observation points most likely have not been visited before the survey and their exact location and condition may not be determined until the sites are visited. For these surveys, planning in detail for each site is rarely practical because unanticipated legal and physical barriers to access of survey locations may change with time. In addition, the preservation of safe working conditions related to traffic or weather in the field often requires on-the-spot changes in survey plans. With or without pre-planning, detailed notes on the survey should be taken at the time of data acquisition. Complete and detailed notes can significantly reduce data reduction errors and are a necessary component of a quality-assurance program. Without such notes, correction of anomalous points would require a repeat occupation of the field site. Guidelines for documenting field acquisition, either in a notebook or in a computer will help to maintain complete and consistent notes and will speed data reduction.

The field survey techniques must accommodate the size and terrain of the survey area and the available instruments. Each survey will be unique and could require some modifications to survey technique. However, the data documentation should be consistent to encourage a systematic reduction process and record of quality control. Documentation starts with familiarizing the survey crew with the area through a site visit, perhaps when establishing base stations, or by examining maps and aerial photographs. Interactive web sites provide a useful tool for examination of the acquisition area. The importance of documentation and its contribution to efficiency is fully realized when a survey is interrupted or when problems in meter drift or locations are discovered after the survey team has left the field. Although GPS systems and meters equipped with computers reduce the need for hard copies, backups should be designed to prevent loss of data in power failures or computer crashes. Ultimately, a backup of essential data should be printed in a format easily interpreted by optical character recognition.

General guidelines that apply to all gravity surveys are as follows:

(1) Preserve safe working conditions. In addition to the possibility of personal injury, an accident of any type can cause delays and lost data far in excess of the time

lost by avoiding accident-prone situations or finding an alternate data-observation site.

(2) Understand the legal requirements for permission to survey on public and private rights-of-way. These can depend on local governments as well as the type of organization doing the survey.

(3) Let authorities know what you are doing and when you plan to be in their area of responsibility. If your base station is near a public building with security concerns, do not show up unannounced at a strange hour with unfamiliar looking equipment.

(4) Establish a measuring procedure that is consistent (the same for all points), practical, timely, and efficient. Standard operating procedures (SOPs) should be developed and followed for the data acquisition and quality control. Each survey can have unique requirements, so the operator should be familiar with general standards, such as the "Standard Guide for Using the Gravity Method for Subsurface Investigation." (ASTM D6430, 1999), and site-specific SOPs. A good measuring procedure will allow efficient flow of data from meter measurement to reduced gravity values with a minimum opportunity for introduction of transcription errors. A spread sheet or data base in a portable computer can allow timely field reduction, but the operator should maintain extensive backups and/or hard copies of the field notes.

(5) Establish a mechanism for complete documentation of the survey data at the time of acquisition. The position of a meter at a measurement station can be preserved nicely with a photograph. The documentation should be complete enough to allow the survey to be repeated by a second survey team, and sufficiently detailed to permit correction of location, time, or other errors without repeating the field acquisition.

3.2 Planning the survey

The survey parameters, including number of points, station separation, and area should be determined prior to field acquisition and should be based on the size and location of the target. For regional surveys in which crustal structure is being investigated, gravity stations may be acquired along roads at a wide spacing, such as 0.5 km. For detailed microgravity surveys, gravity stations may need to be acquired in a regular grid pattern at a tight spacing, such as 1 to 5 m. Ultimately, the project budget will constrain the total number of measurements that can be acquired, and therefore, the station layout should adequately cover the area of interest in the most cost-effective manner. The total width of a survey will ideally extend beyond the influence of the target structure. The depth of the target will determine the data separation. The station separation should be smaller than the depth to the target structure because that separation would be the minimum required to characterize the anomaly. Alignment of the survey lines or grid points at an angle (preferably perpendicular) to the target of interest improves the reliability of resolving an anomaly. In regional surveys, consideration should be given to obtaining some data along lines with very small data separation. Such detailed line data will enhance the interpretation of shallow features and shed light on the sources of noise in the gravity field.

Planning sessions prior to starting the entire survey and each day's survey are helpful. If a new base station is to be used, its location should be chosen and its value determined before field data are collected. A survey is usually made with a two-person work party and the best (and efficient) way is to work out a division of responsibility between the two party members. In a regional survey using a vehicle, it is best to let the driver take readings and the other person navigate (write descriptions, find elevations, plan for future points, etc.). The areas to be surveyed should be inspected by all survey members (on topographic maps or aerial photographs) and a general method of attack should be established. Before beginning the day's survey, plan the area to be covered. Determine which roads are to be used and where to obtain supplemental points not on the road. Do not try to establish the exact location of each point in regional surveys before going into the field because points so chosen may not be suitable for a gravity reading. For example, the location may be difficult to identify on the map or the roads may have changed since the topographic maps were compiled. In regional surveys there is often a temptation to take readings at established bench marks, often designated by a brass disk. These provide surveyed locations and elevations. The down side is that these are often difficult to find, have been moved or have been destroyed, and may not use the same elevation datum as your GPS. In general, a starting area should be chosen and readings should be taken systematically from one side to the other. For example, if a new base station is established near the center of the area, a starting area might be north of the base station and general areas covered in a clockwise fashion.

3.3 Suggested documentation

For regional surveys, each gravity location point is recorded so that it can be reviewed as a point plotted on a topographic map. One way to accomplish this in the field is by saving a waypoint in a GPS instrument. The symbol on the map or the identification of the waypoint should contain a unique survey number and reading number. Later, any questions concerning possible errors in the data location can easily be traced to the originating survey. If maps are being used for elevation control, care should be taken to see that the current topography does not differ significantly from that depicted on available topographic maps. Remember that roads may have changed since the maps were compiled. Contours on older maps can differ from the current ground surface. The elevation and location determined from GPS system may not be affected by surface changes, but if digital versions of maps are used for topographic corrections these changes in topography could affect terrain corrections. Remember that GPS elevations are with respect to the ellipsoid and map elevations are referenced to the geoid. The reduced data using GPS should be labeled ellipsoidal anomalies.

Obtaining the horizontal geographic location using differential GPS with sub-meter precision is adequate for most regional and local microgravity surveys because the effects of latitude variation are less than 0.001 mGal per meter. Sub-meter elevation precision is acceptable for regional surveys, limiting elevation-related errors to less than 0.1 mGal, but would not replace elevation surveys in local or more precise field measurements in microgravity surveys. In microgravity surveys, an elevation survey with a reading resolution within 3 mm and a loop-closure precision within 3 cm/km should be used.

For microgravity surveys, great care should be used in installing and surveying the gravity stations. The stations should be marked in advance of the gravity measurements. A 60 d nail with a stake marker works well as a survey station for a microgravity survey. Localized terrain variations, such as tire ruts, a berm, or a drainage ditch should be avoided, as these features will produce local topographic anomalies that add noise to the data obtained at the microgravity scale. Areas with similar irregular topography should be avoided. The elevation of each station must be obtained with a precision appropriate for the magnitude of the target anomaly and data noise. Caution is needed in taking measurements in and near tall buildings because these values will be affected by the building structure and the effects of construction of the foundation.

In preparing the documentation, the survey documentation contains information common to each station. The survey documentation should include the following:

(1) survey area (quad sheet(s), highways, nearest towns);
(2) party members and responsibilities;
(3) the time standard (standard time, daylight savings time, time zone);
(4) type of instrument used and any information like normal operating temperature that would be characteristic of that particular type of instrument;
(5) instrument serial number;
(6) survey number or unique name; and
(7) date.

The station documentation should include all information specific to individual stations, such as following:

(1) Station number or unique label.
(2) Latitude or relative northing (include source).
(3) Longitude or relative easting (include source).
(4) Elevation. Indicate whether from contours, benchmarks, GPS, differential GPS, or survey techniques. Provide an estimate of uncertainty for each.
(5) Instrument height above the measurement point.
(6) Meter temperature, battery voltage or other factors that affect performance of the meter.
(7) Time (use 24-h clock and note time zone and whether it is daylight savings time).
(8) Meter reading.
(9) Verbal description of the location.
(10) Graphical description of the location (photograph).

3.4 Base station network

A base station that is convenient to the survey area should be established before data are acquired. For regional gravity surveys and for surveys in which the data are to be compared to data from previous or future surveys, the value of the gravity at the base station should be tied into a state or national base-station network. The tie to a state base allows one to reference the data to absolute gravity and to compare your data with data from other surveys in the same area. For local microgravity surveys where a tie into regional data is not necessary, an assumed absolute gravity value can be assigned to the base station. The location of the base station should be accessible at any time of the day. If multiple surveys are to be combined, a common base tie is necessary.

In most surveys, a new base station significantly improves convenience. For example, if the state or national base is far from the area of a survey, a new base may be established to make base visitations more accessible and less time consuming. In order to establish a new base station, a location near the center of the survey area is chosen. The new base should be easily accessed by common roads and should be located at a relatively permanent site. Usually, locations near churches, post offices and court houses are a good choice, but chose a location that does not appear to present a security threat to the facility. Road intersections are not appropriate because traffic at certain times during the day may interfere with the reading and the station location could be altered by subsequent road work. The base station should be located on a flat rock outcrop or concrete slab. The verbal and graphic descriptions of the base station (and all other gravity points) should be adequate to allow someone who is unfamiliar with the area of the survey to locate the base (or other points) to at least within one meter. This means describing the location of the station relative to local permanent landmarks by maintaining reasonable detail in field notes, by taking photographs, or by using GPS with 1-m precision.

The value of gravity at the base station is found in a separate survey in which the state base station and the new base stations are visited two or three times. Each reading should be made carefully to minimize any discrepancy between readings. Because tidal effects and instrument drift will change for each reading, the readings are expected to be slightly different. Great care in these readings is required because the rest of your survey will be affected by this one value and if the base value is in error, then your data will not agree with points obtained on other surveys. Three or more comparisons are needed to demonstrate the precision of the base tie and to obtain a statistical estimate of the value and its precision. For example, seven readings taken alternately at the state base and the new base would give three independent estimates of the value at the new base station and four readings to define meter drift. Provided no errors are detected, the three values may be averaged, resulting in a single estimated value of gravity at the new base. In addition, the precision of the gravity at the new base can be estimated. If the calculated values differ by more than the precision needed in the survey the data should be examined carefully for errors and redone if necessary to give a precision consistent with the meter and survey. A precise base value is important because it is used to define every point in the survey.

3.5 Monitoring meter drift

All relative gravity meters inherently drift, meaning that the measured gravity value changes as a function of time. This drift must be monitored in order to remove it during the data processing. The drift is monitored by making measurements at the established base station throughout the course of the survey. The recommended frequency of base station re-occupations is dependent on the type of gravimeter being used. Older meters often have a non-linear drift over several hours, and therefore it is necessary to re-occupy a base station approximately once per hour to account for the drift. More modern meters have a much more linear drift, and it may be only necessary to acquire a base station reading at the start and end of each day. A good practice is to visit a base before and after any long breaks, such as a lunch break. Because the drift of the gravity meter may vary depending on its temperature, the temperature and the voltage of the battery used to heat the meter should be monitored. A nearly-exhausted battery will not maintain the meter at the proper temperature. The temperature in hot climates should be monitored closely. If the outside temperature approaches the meter temperature, as it might on a hot pavement at noon, the meter's ability to maintain a constant temperature and stable reading is reduced. When the gravity values are corrected for the drift, the drift is assumed to be a linear function of time. The correction for drift is then made to the gravity values taken between base visitations.

Other factors can also affect instrument drift. For example, meters using an unclamped quartz spring, are susceptible to high levels of non-linear drift if they are not kept level during an extended period of time. Therefore, it is advisable to allow this non-linear drift to dissipate after transporting a meter or leaving it off-level for several hours.

Data acquisition at field stations will vary slightly depending on the instrument being used, and, therefore, standard operating procedures and manufacturers' instructions should be followed. However, in all cases, one should ensure that the reading is a repeatable and valid data point for the survey. The meter base should be firmly pressed into the ground and stabilized prior to the reading. If the meter slowly sinks into the ground during the reading (in soft materials such as hot asphalt) the meter will be off-level and produce errors in the data. When using modern gravimeters that integrate many readings over a specified time window to improve the signal-to-noise ratio, one should be aware of noise spikes in the readings from wind gusts or nearby traffic.

It is imperative that the instrument and ambient noise be monitored throughout the survey. The only reliable method to monitor this noise is to re-occupy and re-measure data points throughout the survey. At least 10 percent of the field stations should be re-measured at random times throughout the survey. This will provide a measure of the precision and repeatability of the measurements that accounts for all sources of noise. It is not acceptable to simply make multiple consecutive readings at several field stations or to rely on the real-time noise monitor on modern gravimeters, which will not provide an accurate measure of noise due to base station loop errors, tares, and the effects of transporting the meter between stations.

3.6 Same-day data reduction

Each survey should be plotted and reduced as soon as possible after the survey was taken. The data description forms (or computer files) should be arranged for easy entry into the reduction program, or maintained on a portable computer with immediate backup. In general, the more often the data are rewritten, the more the data can be expected to have transcription errors introduced into the reduction.

After the data have been reduced, visually review an independent plotting of the data and compare that plot with the map used in the field survey. Incorrect locations are easily found in this manner. Also, observe the values of gravity to insure that the gravity varies smoothly. If a value doesn't fit in with the surrounding values, first check the elevation (assuming location errors are eliminated). Then, check the data that have been entered into the computer with the data in the survey notes. If the discrepancy has not been found, reread the point on a later survey.

4 Graphical representation of the anomalous field

4.1 Map scale and implied accuracy

Gravity data are usually presented in a graphical mode as a contour or relief map. The contour map may be color coded to indicate the magnitude of the anomaly. Also, the contours may be shaded to simulate a low angle of illumination or displayed as a surface to provide a three-dimensional visualization of the data (Figure 4.1). Such presentations facilitate visual perception of the shape and trends of the potential field. Color coding and shading have the advantage of minimizing the mental effort needed to convert contour lines and point plots of data to a visual image of the surface. The similarity of this image to the familiar presentation of topography facilitates interpretation and identification of anomalies. With an understanding of how density anomalies contribute to the anomalous field, observed anomalies can be translated into estimates of structures that ultimately lead to more quantitative interpretations of structure.

Often, access to an area is limited and in order to obtain sufficient detail for interpretation, the data are obtained along lines and projected onto straight-line profiles that are approximately parallel to the line. These slices through the gravity field of a study area can be analyzed in more detail and provide interpreted information more quickly than from the larger data set required to cover a two-dimensional area. Whether it is a line of closely spaced data or a section drawn through a two-dimensional map, these one-dimensional profiles generally require that the trend in the anomalies be known before an analysis is started. Applications of two-dimensional models require that the slices be projected onto a line perpendicular to the trend of the structure.

A critical component in preparing a graphical depiction of a three-dimensional surface is the map scale. The map scale controls the ability to compare contours with a point plot of the data and (in data reduction) the map scale defines the ease in examining discrete values. In electronic form, the gravity station locations may be viewed at any scale and overlain onto topography and other spatial data using GIS software. However, when preparing figures in a report or a printed map, the choice of map scale may be dictated by the density of posted values and the use of maps covering adjacent or overlapping areas. This is particularly important when the data for these maps are not available in a compatible electronic format for plotting. Also, the choice of the map scale may be determined by the advantages of direct comparison with published geologic maps, such as the 1:500000 scale state maps or the 7.5 minute quadrangle maps used for more detailed geologic field studies.

Gravity data can be converted to a contour map by many commercial programs. However, an improper choice of contouring techniques can destroy the objectivity in the presentation

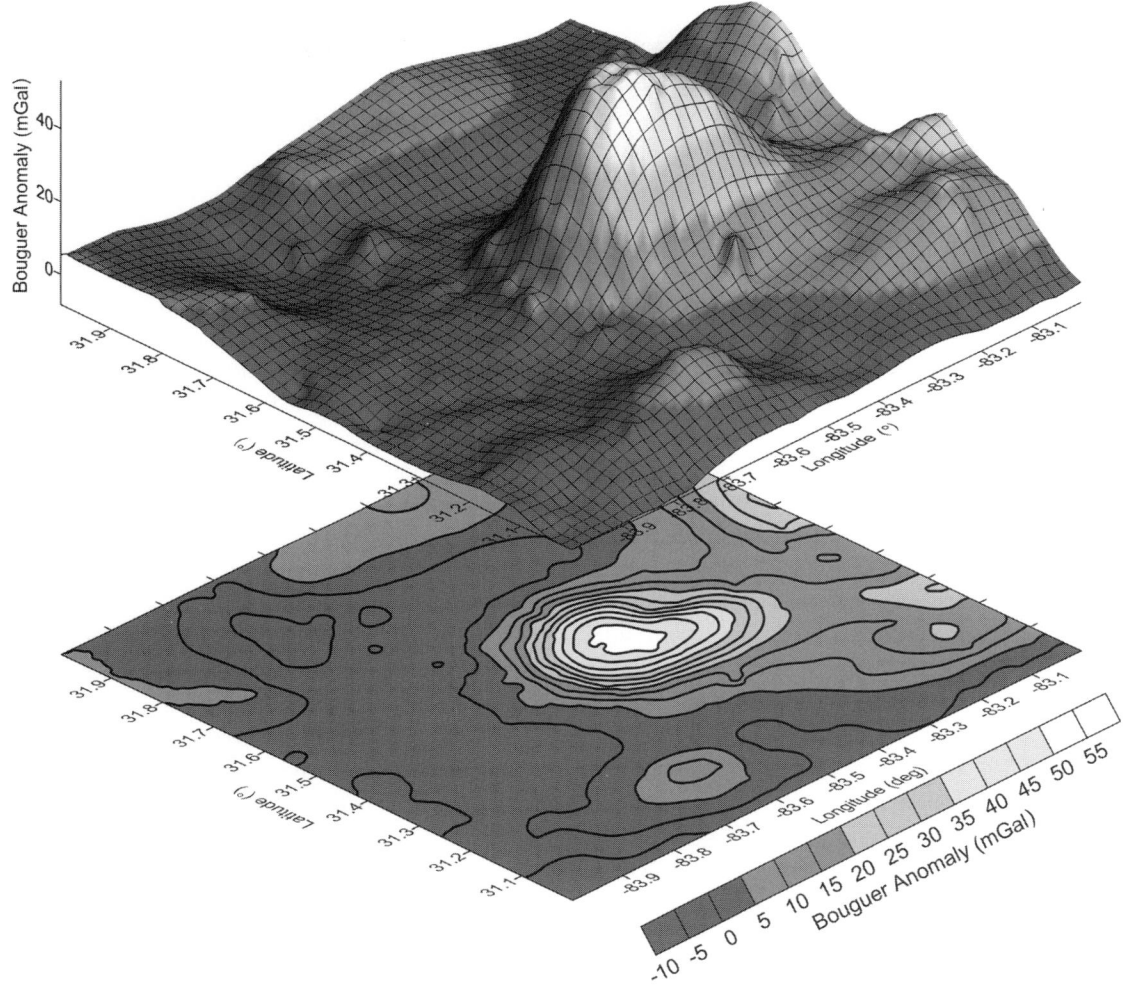

Figure 4.1 Graphical representation of a gravity anomaly in the Coastal Plain of Georgia, USA.

of data and introduce false and misleading anomalies in areas poorly constrained by data. All contouring techniques must use an interpolation method to extrapolate data from observation sites to areas without data coverage. Extrapolation, particularly in those techniques that fit sparse data to flat or curved surfaces, is prone to uncertainty in defining the slope of the surface. When the slope is in error, extrapolation into unconstrained areas can generate large artifacts that distort the existing field and can lead to errors in interpretation. Two examples are worth noting because of their common appearance in irregularly spaced data. The first occurs at the edges of study areas where data do not completely fill the area and data from outside of the study area are lacking. Usually the values near the edge are extrapolated to the edge as unconstrained values. If the slope of the gravity field is used in the extrapolation, large edge-effect anomalies can appear. The contours at the edge of the

Table 4.1 Implied precision of a gravity map versus contour interval.		
Contour Interval	Precision of Map	Data Separation
10 mGal	±4 mGal	10 km
5 mGal	±2 mGal	5 km
2 mGal	±0.8 mGal	2–3 km
1 mGal	±0.4 mGal	1 km
0.01 mGal	±0.004 mGal	1 m

map were inappropriate because the distance to the nearest observation point exceeded that in the center of the map and is biased toward the direction of the data. A second example of contouring artifacts appears in areas where data are obtained at close intervals along lines and used to extrapolate between lines that are separated by distances much greater than the data spacing along the lines. The extrapolation typically uses the slope determined from one line, for example, from the eight nearest points. Because the points are nearly along a line, the determination of the slope perpendicular to the line is poorly constrained and prone to large errors. Artifacts of this type appear as pairs of positive and negative anomalies between lines of data. In this case, the high density of data along a line reduces the independence of the data, and a larger area should have been used in determining the slope. Dense spacing along a line is useful for profile interpretation, but should be used cautiously in defining the two-dimensional character of the field. Some contouring programs first interpolate the data to an evenly spaced grid. When contouring study areas with variable spacing of data, areas of dense data can lose detail and areas of sparse data will appear more detailed than appropriate for the data available. Areas with greater density should be indicated by using an appropriate contour interval, one that preserves the detail, or by some other indication on the map.

In properly prepared contour maps the contour interval should indicate the general precision of the values at any position on the map. For gravity maps, the data separation and the rate at which data values tend to change with distance determine the uncertainty of extrapolated values. Hence, the precision of any arbitrary position is functionally related to the distance of that point from observed gravity data. The range of possible values that a gravity value can take increases with distance from points where the gravity value is known. Table 4.1 gives values of precision for typical gravity data as a function of data separation and an appropriate contour interval for that level of precision.

The uncertainty of values determined by extrapolation to any location on a gravity map can be used to determine an appropriate contour interval from the empirical relation, $d = 2.5/m_{\Delta g}$, where $m_{\Delta g}$ is the standard deviation of the errors of an arbitrary position within the survey area. This relation is constrained by the rate of variation in the density structure near a gravity observation. Areas in which topography is slight and the lateral variation in density is small could decrease the uncertainty for a given station spacing. Note that the precision of the map includes but is not equivalent to the data precision, which should be measured by repeated measurements during data acquisition (see Section 4.3).

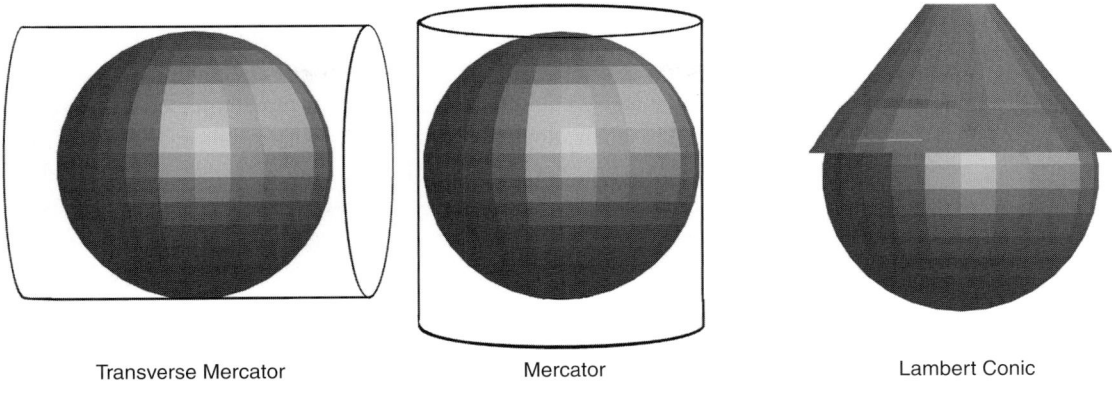

Transverse Mercator   Mercator   Lambert Conic

Figure 4.2   Three commons projections for mapping a spherical surface to a flat surface.

4.2 Map projections

Because the Earth's surface is curved, distances and direction are distorted when projected to a flat surface, such as a computer screen or paper map. The distance between two points and directions change according to the type of projection used. Most local surveys are in areas that are less than a few tens of kilometers and the area can be projected to a plane with negligible distortion. The maximum distortion at a distance of 600 km for projections to a flat surface is on the order of 1.0 km. However, for a distance of 60 km the map distance could be measured with a relative precision of 1 m. Prior to the use of GPS receivers, gravity survey locations were referenced to the geoid, the equipotential surface of the Earth, and would include undulations and distortions that follow the gravity anomalies. For non-GPS location references, not only are there distortions related to the type of projection, but there are also minor distortions related to undulations of the geoid. Satellite-based GPS system locations remove some of these minor irregularities because they are referenced to a geometric system independent of the geoid undulations, but they are not immune from the distortions associated with projection to a flat surface. In practice, differences in the distortion of map scale related to the different projections used to map spherical coordinated onto a rectangular coordinate system are important only for regional surveys covering more than a degree. The effects of geoid undulations are normally obscured by the natural short-wavelength variations in the gravity field.

The more common projections are the Transverse Mercator, Mercator, and Lambert conformal conic (Figure 4.2). The Mercator projection uses the parallels and meridians to form a rectangular grid and is easily plotted. The spherical surface is projected on a cylinder with its axis parallel to the axis of rotation of the Earth. Distances in the north–south direction are accurate, but near the poles, the scale for east–west direction is highly exaggerated. The Mercator projection is widely used for navigation because all straight lines are lines of true bearing or azimuth. The Lambert conformal conic projection projects the spherical surface on a simple cone. The cone intersects the Earth along two standard parallels of latitude on one side of the equator. The meridians are straight lines that meet at the apex of the cone and the parallels are concentric circles. Areas between the standard parallels are slightly reduced and outside the standard parallels they are slightly increased. The scale errors rarely exceed 1 percent. The transverse Mercator projection and the universal Transverse Mercator grid have been adopted for many topographic and geologic maps. The transverse Mercator projection corresponds to projection onto a cylinder wrapped about the Earth, tangent to a central meridian. At the central meridian, the longitude lines are straight but become more distorted with distance from the central meridian. The universal transverse Mercator grid is a rectangular grid overlaid on the projection and has been designed for world use to 84 degrees north latitude. The globe is divided into zones of six degrees of longitude in width. The central meridian of longitude is assigned an arbitrary value of 500000 m to avoid negative numbers.

For station locations determined prior to GPS availability, one should be aware that the accuracy of the position can be affected by factors independent of the projection. In most printed maps the paper can be distorted through shrinkage or by excessive folding and distortion at creases. These distortions may be as high as 3 percent, but they can generally be avoided by using reference points close to the survey point or by avoiding creases in the map. Round-off or truncation of precision in archived data can also contribute to the uncertainty.

4.3 The accuracy of gravity data measurements

The accuracy of gravity measurements in a survey is determined by a combination of systematic errors, which are designated $C_{\Delta g}$, and uncertainties in the measurements, which are designated m. The systematic errors are constant or smoothly varying over the survey area and are not always well defined. They are typically constant for all of the data or for sub-sets of the data. They arise most often from errors in base ties to absolute gravity. Meter-calibration errors can affect the accurate conversion of meter scale readings to gravity values. Tares can also contribute systematic errors in points from a portion of a single survey. Occasionally, different, usually older, surveys will be reduced using a different reduction equation or a tie to an older base station network. Two surveys that are based on different reduction equations or base ties need to be tied together with duplicate measurements. In that way systematic differences can be identified and removed.

The precision of a gravity map, $m_{\Delta g}$, may be estimated from the precision of individual readings and values of gravity that are extrapolated to arbitrary points on the map. The accuracy at any point is the combination of systematic errors and the measure of the precision through the relation

$$M_{\Delta g}^2 = m_{\Delta g}^2 + C_{\Delta g}^2. \tag{4.1}$$

The important component in Eq. (4.1) is precision, the mean square error $m_{\Delta g}^2$. The systematic errors are generally constant over the survey area and their influence amounts to a simple shift in the reference level. A constant shift in gravity values would not influence most interpretations of structure. In contrast, the mean square error is important in establishing confidence in the shape of the anomalies and, hence, can strongly influence interpretation of structural details.

The precision of an arbitrary position on a gravity map is a combination of the effects of the uncertainty associated with individual observation points, m_i, and errors associated with the process of interpolation. As expected, the uncertainty increases with increasing distance from points of measurement. The precision of an individual gravity measurement may be expressed as a combination of contributions from all the factors that have gone into its computation. The most common and significant factors would be combined as in the equation

$$m_i^2 = m_g^2 + \left(\frac{\partial \Delta g}{\partial \phi}\right)^2 m_\phi^2 + \left(\frac{\partial \Delta g}{\partial h}\right)^2 m_h^2 + m_{\Delta g_{\text{terr}}}^2, \tag{4.2}$$

where m_i^2 is the mean square error computed for the ith station. The terms on the right-hand side of the equation are, respectively, the gravity value variance of instrumental measurement, uncertainty in latitude, uncertainty in elevation, and uncertainty in the terrain correction.

The variance of the gravity values $m_g{}^2$ is a function of meter reading precision, drift stability, site conditions, and scale resolution. The reading precision of most modern relative gravimeters is 0.001 mGal, which is typically much better than the uncertainty of the gravity under normal field conditions. Wind, localized site vibrations, distant earthquakes, and soil conditions will all contribute to the variance in the data. Gravity surveys will have a data variance as low as ±0.003 mGal for precision microgravity surveys to ±0.03 mGal for regional surveys. Repeated measurements should be made during different base station loops in a pseudo-random fashion in order to provide a representative measure of data precision. When integrating older surveys, the reading precision of the meter can become important, as some older meters obtained data with precision no better than 0.01 mGal (North American) or 0.06 mGal (Worden educator model).

The uncertainty associated with location is controlled by the latitude term in the equation for theoretical gravity. The rate of change in gravity with latitude is 0.8 mGal/km at 40 degrees north latitude. On most 7.5 minute quadrangle maps the location can be determined easily to within 100 m, or an uncertainty of less than 0.08 mGal. The sub-meter precision of differential GPS systems can reduce this to ±0.0008 mGal.

The uncertainty with elevation is determined from the free air and Bouguer plate corrections, which combined are approximately 0.2 mGal/m. A precision of ±0.3 m, the best that can be expected from spot elevations on a quadrangle map implies a precision of ±0.06 mGal. For general observation points measured from a topographic map or from a standard differential GPS receiver, the elevation precision is seldom better than a couple of meters, an uncertainty of ±0.4 mGal. More precise GPS systems provide elevation precision in the 12 to 18 mm range, which implies a precision of ±0.003 mGal, which is sufficient for most regional and local gravity surveys. The precision of the technique used for elevation control should be established for each survey area and instrument. For precise detection of cavities, a precision of ±0.001 mGal would require an elevation precision of 5 mm and require a precise leveling survey of the gravity stations.

The terrain correction, which is always positive, varies according to the magnitude of the topography close to the station. Values of 0.01 to 0.3 mGal (±0.15 mGal) are found in areas of slight topography, whereas mountainous areas may have changes in value of 5 to 10 mGal over distances of a few kilometers. These terrain effects, when properly applied during the data-reduction sequence, will be removed from the Bouguer data. The terrain correction is only as good as the elevation data (digital elevation model or other topographic data). Digital elevation models are routinely available at one arc-second spacing (approximately 30 m), which is adequate for most regional surveys. However, local and microgravity surveys often require much tighter elevation sampling, especially in areas of uneven terrain. In these cases, a digital elevation model (DEM) with 1/3 arc-second (10 m) spacing or LiDAR data are more appropriate. Variations of topography that are not adequately sampled by the DEMs will contribute to uncertainties in the terrain correction.

4.4 Observational determination of precision

In any survey, it is good practice to repeat measurements at scattered points. A total of 10 percent to 15 percent of the measurements should be repeated for quality control. These duplicate points, which are field stations and not base stations, provide a valuable resource for checking for base tie errors or inconsistencies when independent surveys are merged. Repeat observations allow an estimation of error. Data points are independent when they are determined by a separate and independent determination of meter reading, location, and elevation as done by different observers. The equation that may be used to approximate the uncertainty of individual observations points is

$$m^2_{\Delta g} = \sum_{i=1}^{n} \frac{\left(\Delta g_i - \overline{\Delta g} \right)^2}{(n - v)}, \tag{4.3}$$

where $(n - v)$ is the number of excess readings from sites that were occupied more than once. Equation (4.3) will underestimate the uncertainties in measuring gravity because it is only a measure of the precision in meter reading and data reduction, and does not include the variations of gravity caused by local density anomalies with dimensions less than the data separation. In practice, a truly random sampling of repeat measurements is difficult to obtain. However, measurements should be repeated in several quality-control patterns.

Multiple successive measurements without moving the meter will determine the station-specific noise at a particular time period. This is a good measure of external noise factors such as wind and vibrations that are influencing the measurement. Also, with many older meters multiple readings at a single station will identify reading variations related to back-lash in the adjustment mechanism. Backlash occurs when the screw threads become worn and give different readings when the reading line is approached from opposite directions. If such a variation is discovered, subsequent readings should be made by approaching the zero line always from the same direction.

Multiple measurements of selected stations at different times throughout the survey will incorporate the station-specific noise with errors in base-station ties, errors in meter drift, and errors due to slight differences in meter positioning at a single site relative to local terrain.

4.5 Linear interpolation of gravity data

Once the precision of an individual observation point is determined, the next step in computing the precision of a gravity map is to evaluate the errors introduced by extrapolation to unoccupied points. The loss of precision in extrapolated values is likely to be underestimated because points of difficult access and uncertainty in local density structure or topography are systematically avoided in most field surveys. The computation of an interpolated gravity value, Δg_t, can usually be expressed as the linear weighted sum of individual neighboring gravity data by the equation

$$\Delta g_t = \sum_{i=1}^{n} \alpha_{ti} \Delta g_i,\tag{4.4}$$

where α_{ti} is the weight function for the contribution of the ith point to the extrapolated value, Δg_t. The weight function is constrained by the normalization relation

$$\sum_{i=1}^{n} \alpha_{ti} = 1.\tag{4.5}$$

Weights that are simple functions of the distance, r, from the observation point to the extrapolated point are common and easy to evaluate in equations for precision. Some examples are shown in Figure 4.3.

In Figure 4.3 the weight functions were computed from the following relations:

$$\alpha_{ti} = \frac{k}{r_{ti}}\tag{4.6}$$

$$\alpha_{ti} = \frac{k}{r_{ti}^n}\tag{4.7}$$

$$\alpha_{ti} = \frac{k}{\left(1 + \left(\frac{r_{ti}}{a}\right)\right)^n}\tag{4.8}$$

$$\alpha_{ti} = \frac{k}{1 + \left(\frac{r_{ti}}{a}\right)^n}.\tag{4.9}$$

The value of k is computed according to Eq. (4.5) from the weights of all data included in the extrapolation.

Another example of a linear interpolation method is planar interpolation within triangles (Figure 4.4) with corners that are defined by the observation points. The gravity anomaly covering the map area is assumed to define a plane within the triangular areas defined by

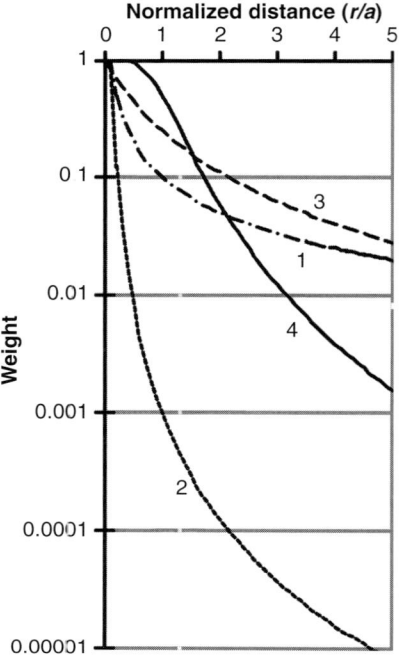

Figure 4.3 Four weight functions for averaging and extrapolating spatially separated data corresponding to Eqs. (4.6) through (4.9). Trace 1 is Eq. (4.7) with $n = 1$, trace 2 is Eq. (4.7) with $n = 3$, trace 3 is Eq. (4.8) with $n = 1$ and $a = 1$, and trace 4 is Eq. (4.9) with $n = 4$ and $a = 1$.

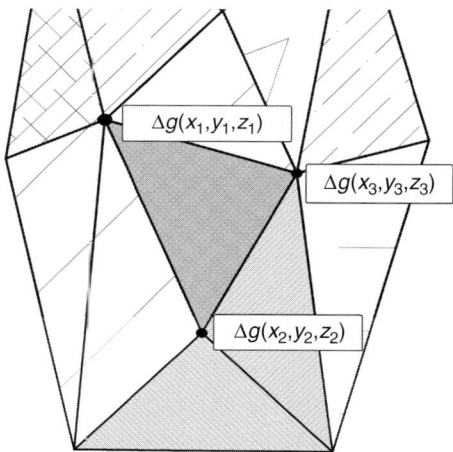

Figure 4.4 Illustration of the use of triangular sub-areas to interpolate gravity data. The lines connecting observation points create the triangles.

convenient data points at the corners. This has the advantage of programmable simplicity, and can be efficient with regularly spaced data points. The uncertainty in the extrapolated position value is not as easy to compute as the value because the uncertainty depends on the rate of change in gravity with distance. The equation of the plane is given by

$$
\begin{aligned}
\Delta g_t = {} & \frac{(x_2 - x)(y_3 - y_2) - (y_2 - y)(x_3 - x_2)}{(x_2 - x_1)(y_3 - y_2) - (y_2 - y_1)(x_3 - x_2)} \Delta g_1 \\
& + \frac{(x_3 - x)(y_1 - y_3) - (y_3 - y)(x_1 - x_3)}{(x_3 - x_2)(y_1 - y_3) - (y_3 - y_2)(x_1 - x_3)} \Delta g_2 \\
& + \frac{(x_1 - x)(y_2 - y_1) - (y_2 - y)(x_2 - x_1)}{(x_1 - x_3)(y_2 - y_1) - (y_1 - y_3)(x_2 - x_1)} \Delta g_3.
\end{aligned}
\tag{4.10}
$$

The equation can be written as a linear weighted average of the gravity values at the corners of the triangles. Equation (4.10) can be written as

$$
\Delta g_t = \alpha_{t1} \Delta g_1 + \alpha_{t2} \Delta g_2 + \alpha_{t3} \Delta g_3,
\tag{4.11}
$$

in order to define the weights for this linear interpolation method. The sum of the weight is always unity.

4.6 Accuracy of linear interpolation

Each observed gravity value may be considered the sum of the true value and an error

$$
\Delta g_i = \Delta g_i^* + \varepsilon_i,
\tag{4.12}
$$

where the asterisk in Δg_i^* denotes the true, but unknown, value of gravity at the ith observation point. Similarly, the error at an extrapolated point, t, may be expressed as the difference between the true value, which may never be known, and the interpolated value by the expression

$$
\varepsilon_t = \Delta g_t - \Delta g_t^*.
\tag{4.13}
$$

The linearly interpolated value of gravity from Eq. (4.4) can be expressed in terms of the true value and its error with the equation

$$
\Delta g_t = \sum_{i=1}^{n} \alpha_{ti} (\Delta g_i^* + \varepsilon_i),
\tag{4.14}
$$

and by substitution of Eq. (4.14) into Eq. (4.13) the error of interpolation becomes

$$
\varepsilon_t = \sum_{i=1}^{n} \alpha_{ti} \Delta g_i^* + \sum_{i=1}^{n} \alpha_{ti} \varepsilon_i - \Delta g_t^*
\tag{4.15}
$$

or

$$
\varepsilon_t = \sum_{i=1}^{n} \alpha_{ti} \Delta g_i^* - \Delta g_t^* + \sum_{i=1}^{n} \alpha_{ti} \varepsilon_i = \varepsilon_t' + \varepsilon_t'',
\tag{4.16}
$$

where

$$\varepsilon_t' = \sum_{i=1}^{n} \alpha_{ti} \Delta g_i^* - \Delta g_t^* \tag{4.17}$$

and

$$\varepsilon_t'' = \sum_{i=1}^{n} \alpha_{ti} \varepsilon_i. \tag{4.18}$$

The first two terms in Eq. (4.16) define ε_t', which computes the errors attributed to the method of interpolation including the inability of the interpolation technique to model the gravity field. The third term in Eq. (4.16) defines ε_t, which represents errors due to the interpolation of the errors. The magnitude of ε_t depends on the method used in interpolation as well as the individual errors of the gravity observations ε_i. The errors of the interpolated values, ε_t, the errors attributed to the method of interpolation ε_t', and the interpolation of the errors ε_t are unknown and cannot be computed individually for any point.

In most cases, the errors may be assumed to be random and normally distributed about a mean of zero. Hence, their variance may be estimated and used to compute the precision of a gravity map. For the precision, we compute values of the mean square error using the expression

$$m_i^2 = \frac{1}{n} \sum_{i=1}^{n} \varepsilon_t^2 = \mathrm{E}\{\varepsilon_i, \varepsilon_i\}, \tag{4.19}$$

where the mean square operator E for a finite length data set is defined by this summation. The m_i in Eq. (4.19) can be recognized as the precision of the observation points. Conditions at some points may warrant different values of the precision, but in typical surveys they can be assumed to be the same for all data points. The mean square error of interpolated values may, similarly, be expressed as

$$m_t^2 = \mathrm{E}\{\varepsilon_t, \varepsilon_t\}. \tag{4.20}$$

The evaluation of m_t starts by considering the square of the sum of the expressions, from Eq. (4.16) for the errors attributed to the method of extrapolation and the extrapolation of the errors,

$$\varepsilon_t \varepsilon_t = \varepsilon_t'^2 + 2\varepsilon_t' \varepsilon_t'' + \varepsilon_t''^2. \tag{4.21}$$

For evaluation, the expressions for ε'_t and ε'' given in Eq. (4.17) and (4.18) are substituted into the three terms on the right-hand side, respectively, as

$$\varepsilon_t' \varepsilon_t' = \sum_{i=1}^{n}\sum_{k=1}^{n} \alpha_{ti}\alpha_{tk} \Delta g_i^* \Delta g_t^* - 2\sum_{i=1}^{n} \alpha_{ti}\Delta g_i^* \Delta g_t^* + \Delta g_t^* \Delta g_t^* \tag{4.22}$$

$$\varepsilon_t' \varepsilon_t'' = \sum_{i=1}^{n}\sum_{k=1}^{n} \alpha_{ti}\alpha_{tk} \Delta g_i^* \varepsilon_k - \sum_{k=1}^{n} \alpha_{tk}\Delta g_t^* \varepsilon_k \tag{4.23}$$

and

$$\varepsilon_t'' \varepsilon_t'' = \sum_{i=1}^{n} \sum_{k=1}^{n} \alpha_{ti} \alpha_{tk} \varepsilon_i \varepsilon_k. \tag{4.24}$$

In order to obtain a mean square error estimate, the errors must be averaged over many interpolated points. Now, compute the mean square value of each of the above three equations. For Eq. (4.24),

$$E\{\varepsilon_t'' \varepsilon_t''\} = \sum_{i=1}^{n} \sum_{k=1}^{n} \alpha_{ti} \alpha_{tk} E\{\varepsilon_i \varepsilon_k\} \tag{4.25}$$

$$E\{\varepsilon_t'' \varepsilon_t''\} = \sum_{i=1}^{n} \alpha_{ti}^2 m_i^2, \tag{4.26}$$

where the terms for $i \neq k$ average to zero for random errors and only the terms for $i = k$ remain and can be recognized as the error estimate for individual data points. For Eq. (4.23),

$$E\{\varepsilon_t' \varepsilon_t''\} = \sum_{i=1}^{n} \sum_{k=1}^{n} \alpha_{ti} \alpha_{tk} E\{\Delta g_i^* \varepsilon_k\} - \sum_{k=1}^{n} \varepsilon_{tk} E\{\Delta g_t^* \varepsilon_k\}, \tag{4.27}$$

the mean square value will go to zero so long as the errors are uncorrelated with variations in the gravity data. Equation (4.22) gives

$$E\{\varepsilon_t' \varepsilon_t'\} = \sum_{i=1}^{n} \sum_{k=1}^{n} \alpha_{ti} \alpha_{tk} E\{\Delta g_i^* \Delta g_k^*\} - 2 \sum_{i=1}^{n} \alpha_{ti} E\{\Delta g_t^* \Delta g_i^*\} + E\{\Delta g_t^* \Delta g_t^*\}. \tag{4.28}$$

In Eq. (4.28) the mean square value of the extrapolated gravity data is the autocovariance function of the gravity field. In general, the autocovariance function of a gravity field measured on a two-dimensional surface is a two-dimensional function. However, if the autocovariance function is a simple distance function that is independent of direction, and the mean value of the gravity data is zero, then the mean square value for the gravity field can be written as

$$E\{\Delta g_i^* \Delta g_j^*\} \approx K_{\Delta g^*}(r_{ij}) \approx K_{\Delta g}(r_{ij}), \tag{4.29}$$

where the second approximation indicates that the autocovariance function of the true values of gravity may be approximated by the autocovariance function of the observed values of gravity when the errors in the gravity field are small relative to the range in gravity values. The decrease in the autocovariance function at short distances is caused by localized variations in rock density and depth of weathering. The heterogeneous density structures of the near surface, such as changes in depth of weathering and localized variations in rock density can contribute changes in gravity values on the order of ± 0.15 mGal at distances shorter than the station spacing. These are the anomalies that are too small to correlate more than a hundred meters and in most surveys may be considered geologic noise.

The variance of the gravity field is the value of the autocovariance function at zero distance. Using the autocovariance function, Eq. (4.28) can be written as

$$E\{\varepsilon_t'\varepsilon_t'\} \approx K_{\Delta g}(0) - 2 \sum_{i=1}^{n} \alpha_{ti} K_{\Delta g}(r_{ti}) + \sum_{i=1}^{n} \sum_{k=1}^{n} \alpha_{ti} \alpha_{tk} K_{\Delta g}(r_{ik}). \tag{4.30}$$

The autocovariance function, $K_{\Delta g}(r)$, may in most cases be determined directly from the data. The ability to determine the autocovariance function depends on the number of data points and the coverage of the map. The map coverage is sufficient when the source and extent of the principal anomalies lie within the map area. The uncertainty in an extrapolated point is then the sum of Eq. (4.30) and Eq. (4.26),

$$m_t^2 = E\{\varepsilon_t \varepsilon_t\} = E\{\varepsilon_t'' \varepsilon_t''\} + E\{\varepsilon_t' \varepsilon_t'\}. \tag{4.31}$$

Hence,

$$m_t^2 = \sum_{i=1}^{n} \alpha_{ti}^2 m_i^2 + K_{\Delta g}(0) - 2 \sum_{i=1}^{n} \alpha_{ti} K_{\Delta g}(r_{ti}) + \sum_{i=1}^{n} \sum_{k=1}^{n} \alpha_{ti} \alpha_{ik} K_{\Delta g}(r_{ik}). \tag{4.32}$$

In a data set where the data are so widely spaced that the autocovariancefunction $K_{\Delta g}(r_{ti})$ is nearly zero for all points except for very few close to an observed gravity value, the error is determined by the variance of the data given by $K_{\Delta g}(0)$ and the uncertainty of the points with the greatest weight

$$m_t^2 \approx K_{\Delta g}(0) + \sum_{i=1}^{n} \alpha_{ti}^2 m_i^2. \tag{4.33}$$

Because the variance of gravity is usually much larger than the uncertainty of individual observations, the variance will dominate the errors. On the other hand, if the data are close enough together to keep the autocovariance function near the maximum value at $K_{\Delta g}(0)$ for most of the data included in the interpolation, then Eq. (4.30) is nearly zero and the error is determined by

$$m_t^2 \approx \sum_{i=1}^{n} \alpha_{ti}^2 m_i^2, \tag{4.34}$$

which is the uncertainty in the data. This is a trivial case because the autocovariance function can be near the maximum only at distances within which the value of gravity changes very slowly.

4.7 Optimal linear interpolation

The error in interpolation to a single point, given in Eq. (4.30), is determined by the relative distances to the observed points and the autocovariance function at those distances. The values of the weights, in general, may be arbitrary to the extent that they can be defined by some function of distance as in Eq. (4.6) through Eq. (4.9). In general, the weights decrease with increased distance in order to give the extrapolated point a value close to the value of the immediate neighbors. However, the autocovariance function contains information about the increases in uncertainty with increases in distance that can be used to find weights that minimize the uncertainty in linear interpolated values. Starting with Eq. (4.30), we seek values of the weights that minimize the error. In Eq. (4.30) the condition for a minimum is that the derivative with respect to each weight is set to zero to give

$$\frac{\partial}{\partial \alpha_{ti}} E\{\varepsilon_t' \varepsilon_t'\} = -2K_{\Delta g}(r_{ti}) + 2\sum_{k=1}^{n} \alpha_{tk} K_{\Delta g}(r_{ik}) = 0, \tag{4.35}$$

or

$$K_{\Delta g}(r_{ti}) = 2\sum_{k=1}^{n} \alpha_{tk} K_{\Delta g}(r_{ik}). \tag{4.36}$$

This set of n linear equations can then be solved for the values of the weights α_{ti} that minimize the extrapolation error for n points used in the interpolation. The solution is

$$\alpha_{tk} = \sum_{i=1}^{n} K_{\Delta g}^{-1}(r_{ik}) K_{\Delta g}(r_{ti}), \tag{4.37}$$

where the exponent (-1) indicates the elements of the inverse of the matrix of autocovariance values at the distances r_{ik}. Inserting Eq. (4.37) into Eq. (4.4) gives the expression for the optimal linear interpolation

$$\Delta g_t = \sum_{k=1}^{n} \alpha_{tk} \Delta g_k = \sum_{i=1}^{n} \sum_{k=1}^{n} K_{\Delta g}^{-1}(r_{ik}) K_{\Delta g}(r_{ti}) \Delta g_k. \tag{4.38}$$

Equation (4.38) shows that in order to obtain a minimum error in linear interpolation, the statistical properties of the gravity data in terms of the autocovariance function must be known. An interpolation with the minimum possible error can be obtained by substituting the solution for the weights into Eq. (4.38). The resulting minimum error will be given by Eq. (4.32). In inverse theory, the expression for the error of interpolation using the least squares solution for the weights is also the maximum likelihood solution when the errors are normally distributed. Substitution of the optimal solution into Eq. (4.30),

$$E\{\varepsilon_t' \varepsilon_t'\} = K_{\Delta g}(0) - 2\sum_{i-1}^{n} \alpha_{ti} K_{\Delta g}(r_{ti}) + \sum_{i=1}^{n} \sum_{k=1}^{n} \alpha_{ti} \alpha_{tk} K_{\Delta g}(r_{ik}) \tag{4.39}$$

gives

$$E\{\varepsilon_t' \varepsilon_t'\} = K_{\Delta g}(0) - 2\sum_{i=1}^{n} \sum_{k=1}^{n} K_{\Delta g}^{-1}(r_{ik}) K_{\Delta g}(r_{ti}) K_{\Delta g}(r_{tk})$$
$$+ \sum_{i=1}^{n} \sum_{j=1}^{n} \sum_{k=1}^{n} \sum_{l=1}^{n} K_{\Delta g}^{-1}(r_{ik}) K_{\Delta g}(r_{ti}) K_{\Delta g}^{-1}(r_{jl}) K_{\Delta g}(r_{tj}) K_{\Delta g}(r_{tl}). \tag{4.40}$$

Equation (4.40) can be further simplified to

$$E\{\varepsilon_t'\varepsilon_t'\} = K_{\Delta g}(0) - \sum_{i=1}^{n}\sum_{k=1}^{n} K_{\Delta g}^{-1}(r_{ik})K_{\Delta g}(r_{ti})K_{\Delta g}(r_{tk}). \tag{4.41}$$

The optimal linear interpolation utilizes the information contained within the autocovariance function to determine the uncertainty in extrapolated data. In many applications the extrapolated value of gravity is obtained by fitting a more complex surface to values found within a given distance. Techniques for fitting these surfaces are given in Chapter 5. Some examples are surfaces defined by two-dimensional polynomials, many orthogonal function sets and the Fourier transform. Gravity values extrapolated by these and other techniques can define a surface but often suffer from unrealistic extrapolations introduced by inappropriate distributions of data. One such inappropriate distribution is data clustered along a line, a situation that poorly defines the slope of a surface perpendicular to the line. Consequently, such extrapolations will have uncertainties greater than those associated with the simpler linear extrapolation.

4.8 Accuracy of the gravity gradient

The error in the difference between two points and the error of the gradient can be computed in a way similar to that used to derive Eq. (4.32). The error in the difference of two points can be written according to Eq. (4.13) as

$$\varepsilon_{dif} = \varepsilon_t - \varepsilon_s = (\Delta g_t - \Delta g_t^*) - (\Delta g_s - \Delta g_s^*). \tag{4.42}$$

The mean square value of the error is

$$E\{\varepsilon_t - \varepsilon_s, \varepsilon_t - \varepsilon_s\} = m_{\text{dif}}^2 = m_t^2 + m_s^2 - 2E\{\varepsilon_t, \varepsilon_s\}, \tag{4.43}$$

where m_t^2 and m_s^2 are obtained from Eq. (4.32) and $E\{\varepsilon_t,\varepsilon_s\}$ is the covariance of the errors of the interpolated values. A procedure similar to the one used to derive Eq. (4.32) is used to evaluate the covariance of the errors, giving

$$m_{\text{dif}}^2 = m_t^2 + m_s^2 - 2 \left[\begin{array}{l} \displaystyle\sum_{i=1}^{n}\alpha_{ti}\alpha_{si}m_i^2 + K_{\Delta g}(r_{st}) - \sum_{k=1}^{n}\alpha_{sk}K_{\Delta g}(r_{kt}) \\ \displaystyle -\sum_{i=1}^{n}\alpha_{ti}K_{\Delta g}(r_{si}) + \sum_{i=1}^{n}\sum_{k=1}^{n}\alpha_{ti}\alpha_{sk}K_{\Delta g}(r_{ik}) \end{array} \right]. \tag{4.44}$$

The correlation of the errors of the interpolated values serves to improve the precision of a difference by averaging or smoothing the gravity field over a larger area. If the errors of interpolation are uncorrelated, the error for a difference is greater than that for a single value.

4.9 Precision of a gravity map

Equation (4.32) alone, or with Eq. (4.40) provide the tools needed to evaluate the precision of a gravity map. The involvement of the autocovariance function in the solution shows that the precision is a function of distance from existing points. Consequently, a computation of precision at randomly selected points would not be expected to give a smooth estimate of precision. A more realistic measure of average precision would be obtained by computing the precision at a set of points that represent maximum distances from existing data points. A computational technique that may be used in place of finding points of maximum separation from existing points is to successively remove one point and compare the observed value to the interpolated value at that position. This technique is appropriate when the data are uniformly spaced. Data that are obtained as closely spaced points along selected lines would bias the precision estimate toward points near the line and not the value interpolated between the lines. The resulting set of precision values can be contoured as an estimate of the distribution of precision in the gravity map.

4.10 The correlation and covariance functions

The covariance function provides a quantitative comparison of two functions. In one dimension for two continuous functions the covariance function is

$$at = C(\tau) = \int_{-\infty}^{\infty} g(t)h^*(t + \tau)dt, \tag{4.45}$$

where the * on the function h indicates that in general the complex conjugate should be taken. The functions g and h are assumed to have the mean removed. In the correlation function, g and h are also normalized. Because gravity data are real, the equations for correlation and covariance would make the complex conjugate irrelevant. The τ is the shift distance, the displacement of g from h in the computation of the correlation function.

When the functions g and h are different functions, Eq. (4.45) defines the cross-correlation function. When the functions g and h are the same function, Eq. (4.45) gives the autocovariance function

$$a(\tau) = \int_{-\infty}^{\infty} g(t)g^*(t + \tau)dt. \tag{4.46}$$

In the autocovariance function it can be seen by symmetry that

$$a(\tau) = a(-\tau) \tag{4.47}$$

and that a maximum value occurs at $\tau = 0$. At zero shifts all values will have the same sign and their products will all be positive. When the functions are shifted some distance τ some of the products will be negative, reducing the value. For a random function, a slight shift leads to a random distribution of products and the autocovariance function of a random function goes to zero quickly. The autocovariance function for random data is the delta function with the magnitude of the power of the trace.

In order for the covariance functions to be finite, the functions g and h must have finite energy. Hence, a condition for using Eq. (4.45) or Eq. (4.46) in order to compute the covariance functions is that the integrals

$$\int_{-\infty}^{\infty} g^2(t)dt \quad \text{and} \quad \int_{-\infty}^{\infty} h^2(t)dt \tag{4.48}$$

are finite constants. Gravity data can satisfy this condition only when large anomalies are isolated in a regional field that is approximately zero.

More often, the gravity field resembles a stationary function, one with properties that do not change significantly with position. For stationary data, the covariance functions represent measures of similarity between functions per unit length of the data used in the computation. For stationary data the covariance function is computed from the limit

$$C(\tau) = \lim_{T \to \infty} \frac{1}{T} \int_{-T/2}^{t/2} g(t)h^*(t+\tau)dt. \tag{4.49}$$

Real gravity data are sampled at unevenly spaced discrete locations. Also, the typical survey area is limited in extent and rarely would one expect the statistical properties of limited survey areas to be similar to the statistical properties of adjacent areas. Consequently, the evaluation of the integral is more conveniently computed as a summation and the limit is rarely achieved with a limited set of gravity values. The formulation of covariance for discrete and finite data is a more practical way to characterize the statistics of a survey area. The autocovariance, Eq. (4.46) for evenly spaced gravity data along a line is written as

$$K(\tau) = \frac{1}{n} \sum_{i=1}^{n} (g_i - \mu_g)(h_{i+\tau} - \mu_h), \tag{4.50}$$

where the μ are the mean values of the functions g and h. From Eq. (4.50) the variance of g can be written as

$$K(0) = \frac{1}{n} \sum_{i=1}^{n} (g_i - \mu_g)^2, \tag{4.51}$$

where the variance is the autocovariance function with zero shift. In statistics, the value of $n - 1$ is used instead of n, arguing that there is one less independent value when the mean is computed from the data. In general, the $n - 1$ is changed to n when the mean is known a priori rather than computed from the data. In any case, where n is small enough to make a difference, there are insufficient data to estimate the variance. With gravity data, we generally believe the mean should be zero, or set to zero by removing a regional trend, but recognize that the limitations of the survey area cannot guarantee a zero mean and, if not accounted for, non-zero means can introduce unreasonable autocovariance values. For gravity data in which the mean may be non-zero the variance should be computed from Eq. (4.51) instead of Eq. (4.19) and the mean is computed from

$$\mu_x = \frac{1}{n} \sum_{i=1}^{n} x_i,$$ (4.52)

where the same n points should be used to compute the mean and the variance. Also, for a valid representation of the gravity field, these n points should be uniformly distributed over the study area.

For two-dimensional data, the autocovariance of a function x with its mean removed is in general a two-dimensional function that can be written as

$$a(\tau_x, \tau_y) = \int_{x=-\infty}^{\infty} \int_{y=-\infty}^{\infty} x(x, y)x(x + \tau_x, y + \tau_y)dxdy.$$ (4.53)

The two-dimensional autocovariance function can be useful in quantifying regional trends in structures. Along the direction of the trend, gravity data can be interpolated to greater distances because anomalies change more slowly than they do perpendicular to the trend. The two-dimensional autocovariance functions will be elongated in the direction of trends.

One way to overcome the problem of a limited size of the study area is to assume that the anomalies are cyclic. In effect the size of the study area is expanded by placing copies of the data end to end. Usually, the data are smoothed to a mean at the edges or padded with extra points to minimize discontinuities. The covariance function then is the sum

$$K(\tau) = \frac{1}{n} \sum_{i=1}^{n-\tau} (g_i - \mu_g)(h_{i+\tau} - \mu_h) + \frac{1}{n} \sum_{i=n-\tau+1}^{n} (g_i - \mu_g)(h_{i-n+\tau} - \mu_h),$$ (4.54)

where the second term matches the first portion of the function h to the last part of function g.

Equation (4.54) is not as efficient for large data sets as an equivalent computation using the fast Fourier transform. The Fourier transform implicitly assumes that functions are cyclic. The cross covariance function for the functions g and h is the convolution of g with the reverse (points taken backward) of h. In the Fourier transform domain, the convolution is a product. The Fourier transform of the covariance function is thus the product of the transforms of g and the transform of the reverse of h. The product in the transforms is

$$g(k)h^*(k) = \sum_{i=-\infty}^{\infty} a(\tau)e^{-ik\tau}d\tau,$$ (4.55)

where $h^*(k)$ is the complex conjugate of the transform of h, which has the effect of computing the reverse of the function h. The essence of the procedure for computing the correlation function is to:

(1) Transform both functions to get $g(k)$ and $h(k)$.
(2) Take the product of $g(k)$ and the complex conjugate of $h(k)$ to get $g(k)h^*(k)$.

(3) Compute the inverse transform of the product in (2) to get the autocovariance function $a(\tau)$.

(4) In order to obtain the non-cyclic autocovariance function using the Fourier transform, the data need to be smoothed to a mean value and padded with enough points to prevent overlap.

4.11 Computation of the autocovariance function

For convenience, the autocovariance is often assumed to be a function of distance independent of azimuth. This assumption works well for regional studies in which the regional trends in the gravity data are limited or randomly distributed. In survey areas where a single regional trend dominates, a two-dimensional autocovariance function should be used. For irregularly spaced data, the covariance function for gravity data may be estimated by computing the variance for all pairs of gravity values at a given separation

$$K(r_j) = \frac{1}{n(r_j)} \sum_{k=1}^{n(r_j)} (x_k - \mu_x)(x_{k+j} - \mu_x), \tag{4.56}$$

where the number of points corresponds to only those pairs of points separated by the distance r_j and the mean value is the mean value for this sub-set of pairs of points. The j identifies the relative shift in position number of a point at that distance from the kth value. One difficulty with this method is that the uncertainty and scatter in the estimates of the covariance function can be large. At close distances the gravity values are large and many values are required to obtain a reasonable estimate of the covariance function. At distances that are approaching the limits of the survey, the choice of pairs is not representative of the entire data set, because they may be restricted to points only at the most distant corners of the survey.

The covariance function of a difference is an alternative method for computing the autocovariance function. This method works well for short distances but is also limited by the restriction of computation to values approaching the limits of the survey area at large distances. The covariance function for a difference can be expressed from Eq. (4.56) as

$$K_{\text{dif}}(r_j) = \frac{1}{n(r_j)} \sum_{k=1}^{n(r_j)} \left[(x_k - \mu_x) - (x_{k+j} - \mu_x) \right]^2, \tag{4.57}$$

or on expansion

$$K_{\text{dif}}(r_j) = \frac{1}{n(r_j)} \sum_{k=1}^{n(r_j)} (x_k - \mu_x)^2 + \frac{1}{n(r_j)} \sum_{k=1}^{n(r_j)} (x_{k+j} - \mu_x)^2$$
$$- \frac{2}{n(r_j)} \sum_{k=1}^{n(r_j)} (x_k - \mu_x)(x_{k+J} - \mu_x). \tag{4.58}$$

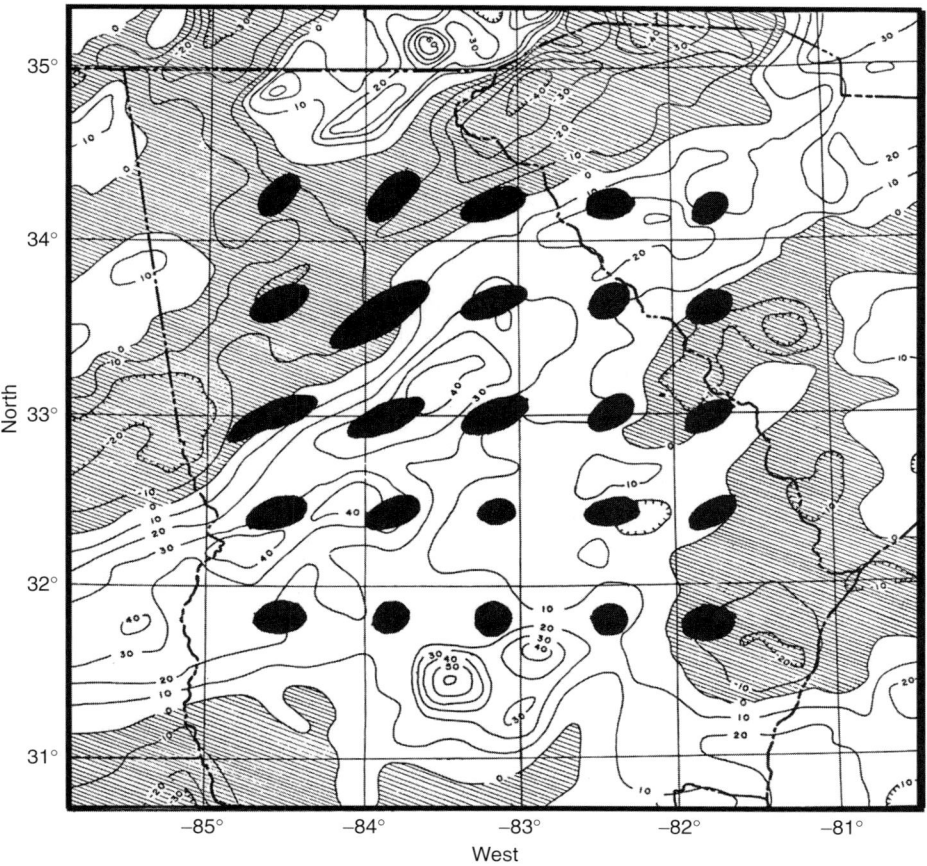

North

Example of application of autocovariance function to bring out variations in data trend of free air gravity anomalies. The ellipses are the half-height of the autocovariance showing trends and correlation distance.

The cross-term is the covariance function and the first two terms are the variance, giving the relation

$$K(r_j) = K(0) - \frac{1}{2} K_{\text{dif}}(r_j). \tag{4.59}$$

The differences are generally small numbers and can be easily computed directly from the data at short distances.

A convenient way to quantify the autocovariance function for gravity data is to fit observed relations to the equation

$$K(r) = K(0) \left(1 - \frac{1}{2} \left(\frac{r}{r_0} \right)^2 \right), \tag{4.60}$$

where r_0 is the correlation distance. When structures are limited to the near surface and their mean is zero, the autocovariance distance will reflect the dominant depth to the causative structures. On the other hand, if the autocovariance distance systematically increases with increased area of study, the anomalies may exhibit scale invariance.

For regularly spaced data, the autocovariance function can be computed from isolated portions of the data set by using the autocovariance expression or the Fourier transform method. When looking at an isolated portion of the data set, the regional component of the field should be removed so that the sub-set has a zero mean and the anomalies being examined are dimensionally smaller than the isolated portion of the map. Figure 4.5 shows one such application to free air gravity anomalies covering the Southern Appalachians and Atlantic Coastal Plane. Where contours demonstrate a trend, the half-height of the autocovariance function is elongated parallel to the structural trend. Areas of more random anomalies have a shorter half-height and do not show a prominent alignment.

5 Manipulation of the gravity field

5.1 Objective of gravity field manipulation

The gravity field may be displayed as discrete points corresponding to points of observation, as an extrapolation of these points to a regularly spaced grid, and as an interpretation in the form of a contour map. Each representation has intrinsic advantages and deficiencies. Although the distribution of observations in the field is rarely in a form that is convenient for an efficient application of computer-based analysis techniques, discrete point locations offer the most precise and honest representation of the data. Also, discrete point locations minimize the introduction of artifacts often associated with extrapolations in graphical presentations of gravity data. The extrapolation of gravity values to a regularly spaced grid simplifies the application of computer programs for analysis; but, the precision of the interpolated value at individual grid point locations can vary widely over the map area. The precision will be low, for example, in areas where the data density is low and where the data distribution is poorly suited for the chosen extrapolation technique. The precision will be high in areas where multiple observations define the value at the grid point location. A contoured presentation displays the data in a form that is easy to visualize. However, a contoured presentation always adds some element of interpretation to the field observations. Contouring programs require assumptions concerning the smoothness of the field in order to extrapolate values to areas without field observations. If the contouring technique, for example, fits a plane to the closest points, then small errors in a tight cluster of data could result in a steep slope to the plane. Extrapolating this plane to adjacent areas without supporting data could lead to large deviations from the actual field value, a common problem with some older contouring programs. Also, contouring programs rely on spatial gridding algorithms such as minimum curvature and kriging, each with their own benefits and limitations of how accurately they honor the data and how they interpolate values. The manual contouring of many older maps by skilled draftsmen often provides a more pleasing and realistic representation of the gravity field. This improved appearance is a consequence of artistic license and preconceived interpretations concerning the cause of the anomalous field. The interpreter may have included information concerning the topography, the density structure, and the geologic units underlying the map area in order to guide the drawing of the details in the contours, or just adjusted the contours to make it look nice. Hence, contours drawn by hand in many older maps can contain more (or less) information than provided by the raw data.

Given a gravity field with the above caveats, the next task is to interpret the anomalous field, generally in terms of the causative density or geologic structure. Such interpretations require that the anomaly of interest be isolated from other anomalies. If one is lucky, 90 percent of the anomalous gravity field of interest will be within the map area and contributions from other density structures or trends in the data will be less than 10 percent of the target anomaly. Unfortunately, such isolated anomalies are rare and one must resort to other methods to isolate the target anomaly. The objective of gravity field manipulation is to transform a gravity field to a form that facilitates the identification and interpretation of the targeted density structure.

5.2 Anomalies: regional, local, and noise

A particular objective in the manipulation of the gravity data is its separation into regional and local anomalies, and the removal of noise. The definition of a "regional" or a "local" anomaly depends on the application. The methods available to facilitate their separation are numerous, perhaps, because many techniques are designed to fit specific gravity anomalies. Also, perhaps, many techniques exist because the separation of regional and local anomalies is limited by the non-uniqueness of gravity data interpretation and the various methods are introduced to force a preferred interpretation on the data. The most common techniques and the methods needed to design new techniques are presented in this chapter.

Regional anomalies, local anomalies and noise may be defined as follows:

"**Regional**" anomalies in a map of finite area are derived from structures that either cover an area much larger than the map or exist outside the map area. They are unwanted because their contribution to the map area cannot be defined by structures within the map area and because they interfere with the interpretation of smaller anomalies derived from the structures of interest.

"**Local**" anomalies are those anomalies that relate directly to the structure or feature of interest. The local anomalies are usually the objective of the survey and are generally contained within the map area.

"**Noise**" usually appears as short-length anomalies unrelated to local anomalies. Noise may be caused by density structures covering areas that are small compared to the data spacing, or they may be random measurement errors in the determination of individual points.

Successful separation of data into regional and local fields, or the removal of noise, is an art in which techniques of varying complexity and sophistication can be applied. In essence, geophysicists with experience in gravity data modeling could easily sketch a residual field that would lead to a model that would be more acceptable, in the geological sense, than most numerical schemes. The reason for this disparity is that a successful separation depends strongly on the preconceived interpretation of the geologic structure. However, if the total field is modeled, then the pattern of density anomalies from the total field should be roughly the same as from the combination of the residual and regional modeling techniques so long as similar constraints are used to restrict the inherent non-uniqueness of potential data. The

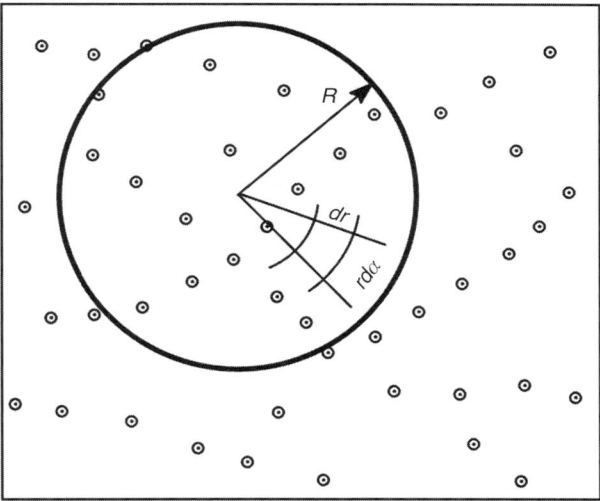

Figure 5.1 Geometrical arrangements for integration over a circular area.

differences in models will depend on the degree of inclusion or exclusion of the regional anomaly in the local anomaly, a separation that cannot be improved without other sources of information on the structure. The difficulty inherent in the separation of regional and local contributions to the anomalous field implies that the interpretation of a local anomaly cannot be considered complete without an accompanying description of the regional field that was removed. As a resolution to the ambiguity of the separation of regional and local anomalies, continuation (described in section 5.3) has been proposed as a standard.

5.3 Smoothing

Much of the noise in gravity data can be removed by smoothing because it is predominantly composed of short-wavelength anomalies and uncorrelated between reading stations. In its simplest form, the smoothing of a gravity field is a moving average. For example, as shown in Figure 5.1, all the points within a circle of radius R centered on an arbitrary point could be averaged and the resulting mean value of gravity assigned to the origin of the circle. If the radius, R, is too small, the average could contain few or no values and would not remove noise. If R is set to include an average of about four observations, then the mean value could reduce the uncorrelated scatter and variations among neighboring values by a factor of two. If R is larger by a factor of 5 or more than the size of the target anomaly, the mean value may be used as a regional value. A good guideline for defining a regional average is to choose the radius of smoothing, R, to be two to three times the expected depth of the anomalous structure or equivalently two to three times the largest significant wavelength of the anomaly to be isolated from regional trends. Although this choice will not separate gradual changes in the target anomalies, it will help find sharp boundaries and

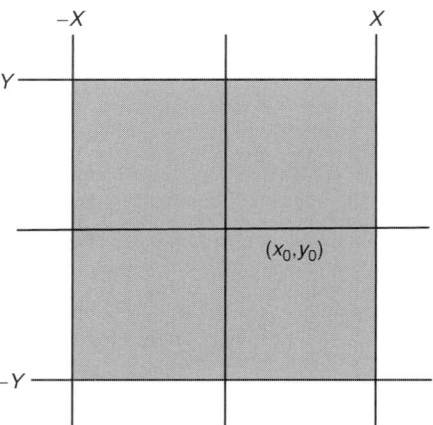

Figure 5.2 Geometrical arrangements for integrating over a rectangular area.

contacts. Care should be used, however, not to over-apply smoothing, especially in local surveys where discontinuities and shorter-wavelength anomalies are of interest.

Smoothing may be expressed as an integral, where the integral is taken over the two-dimensional expression for the gravity field on a surface. In cylindrical coordinates, the origin may be placed at the location of the desired value. First, consider the integration with respect to the azimuth, α, and define the mean value as a function of the radius r as

$$\overline{\Delta g}\,(r) = \frac{1}{2\pi} \int\limits_{0}^{2\pi} \Delta g\,(r, \alpha)\,d\alpha, \tag{5.1}$$

where 2π is the normalizing factor for the integral. The normalizing factor is the value of the integral when a constant value of one is substituted for the value of gravity. For continuous data the gravity values are assumed to be uniformly distributed around the point, but for discrete values, the gravity anomalies should be weighted in proportion to their angle of influence.

In order to find the average value of gravity in a circular area of radius R centered at the origin, the mean values as a function of radius are integrated from the origin to the distance R,

$$\overline{\Delta g} = \frac{2}{R^2} \int\limits_{0}^{R} \overline{\Delta g}\,(r)\,r\,dr, \tag{5.2}$$

where $2/R^2$ is the normalizing factor. In general, the value of gravity cannot be expressed as a simple mathematical function and the integral must be evaluated by using numerical integration. Although, over small areas the gravity may be approximated by simple two-dimensional functions and then evaluated as an approximation in closed form.

For computer applications, a rectangular coordinate system is more convenient (Figure 5.2). In this case, the integral for the mean value of gravity at a central point

(x_0, y_0) may be expressed as

$$\overline{\Delta g}\,(x_0, y_0) = \frac{1}{4XY} \int\limits_{-X}^{X} \int\limits_{-Y}^{Y} \Delta g\,(x, y)\,dx\,dy, \tag{5.3}$$

where the integration is over a square of sides $2X$ and $2Y$, centered at (x_0, y_0). For data interpolated to a regularly spaced grid, each point of the grid represents an average value for the area defined by the grid spacing. Hence, for most averaging techniques the gravity anomaly within a grid increment may be considered to be a constant with the value assigned to the grid point. This assumption is justified by the observation that the process of extrapolating data to a grid usually includes the choice of a grid increment that is short compared to the half-width of the autocovariance function for the gravity data. Hence, the errors in extrapolating between grid points are small relative to the variance in the gravity data. In such a grid, the integral may be replaced by a summation

$$\overline{\Delta g}\,(x_0, y_0) \approx \frac{1}{(2M - 1)\,\Delta x\,(2N - 1)\,\Delta y} \sum_{i=-M}^{M} \sum_{j=-N}^{N} \Delta g\,(x_i, y_j)\,\Delta x \Delta y, \tag{5.4}$$

where the summation is extended over a rectangular area of length $(2M - 1)\,\Delta x$ and width $(2N - 1)\,\Delta y$. The area of averaging can be modified to approximate a circle or other shape such as an ellipse with its long axis oriented parallel to the regional structure. A directional variation in the smoothing area might be justified by examination of the two-dimensional autocovariance function. Placing the long axis of an elliptical smoothing area parallel to the regional structure would allow smoothing of variations in the gravity field in proportion to the rate of change in the anomalies parallel and perpendicular to the regional trend. In such regional trends, the target anomalies are likely to be longer in a direction parallel to the regional structure and a larger smoothing distance parallel to the trend would be needed to avoid distortion of the target anomalies perpendicular to the trend.

An effective technique for retaining detail in closely spaced data and including data in areas of sparse coverage is to decrease the influence (the weight) of more distant points in computing an average. Such averages will be more strongly dependent on closer stations. The mean value of gravity becomes a weighted average where, for example, values at increased distance are given reduced weight. A weight function that is a function of distance, $\alpha(r_j)$, may be introduced into the computation of the average to give a weighted average in the form

$$\overline{\Delta g} = \sum_{j=1}^{n} \alpha(r_j)\overline{\Delta g}(r_j), \tag{5.5}$$

where

$$\sum_{j=1}^{n} \alpha(r_j) = 1.0. \tag{5.6}$$

The constraint placed on the weight function is that the sum of all weights must be unity. Without this constraint the average would be multiplied by an unwanted or undetermined scale factor. Equation (5.5) is a general form of Eq. (5.2) where in Eq. (5.2) the weight function is $r/2R$.

Weighted averages of gravity data can cover a wide range of applications. In their most general form, weighted averages not only describe the common applications to smoothing, but also the computation of directional derivatives, upward continuation, edge detection and other filters that will be considered separately. In particular, smoothing applications of a weighted average are covered under methods of extrapolation of gravity data to a point.

5.4 Examination of a three-point smoothing operator

The summation of three weighted points is a simple example of a weighted average that is useful because its effect can be quantified and it is easy to apply to regularly spaced data. These smoothing filters can be applied to two-dimensional gridded data by sequentially applying the filter in each direction. The result is a smoothing effect that is symmetric in two dimensions. Also, longer filters can often be related to multiple applications of these simple filters. More complex filters can be constructed to accommodate trends in the data, but other techniques like the Fourier transform may prove more practical and appropriate.

The three-point average generates a new value at a defined position, usually at its center, given values at two adjacent points. The three-point average may be written in general form as

$$\overline{\Delta g_0} = \frac{1}{4}\left[4\Delta g_0 + 2s\left(\Delta g_{-1} - 2\Delta g_0 + \Delta g_{+1}\right)\right]. \tag{5.7}$$

Specific values of s may be used to yield different effects. For example, for $s = 0.5$

$$\overline{\Delta g_0} = \frac{1}{4}\left[\left(\Delta g_{-1} + 2\Delta g_0 + \Delta g_{+1}\right)\right] \tag{5.8}$$

for $s = -0.5$

$$\overline{\Delta g_0} = \frac{1}{4}\left[\left(-\Delta g_{-1} + 6\Delta g_0 - \Delta g_{+1}\right)\right], \tag{5.9}$$

and for $s = 2/3$,

$$\overline{\Delta g_0} = \frac{1}{3}\left[\left(\Delta g_{-1} + \Delta g_0 + \Delta g_{+1}\right)\right]. \tag{5.10}$$

Equation (5.10) for $s = 2/3$ is the equally weighted average of three values. The averaging equation for $s = 0.5$ gives a distance-weighted average useful for attenuating noise that does not correlate between values. The equation for $s = -0.5$ has the opposite effect because it amplifies differences between adjacent values. The effects of these filters are more easily understood when examined in the wavenumber domain through the use of the Fourier transform. To examine the response of this smoothing operator, the gravity field may be assumed to consist of a single wave of the form

$$\Delta g = A_0 e^{-ikx}, \tag{5.11}$$

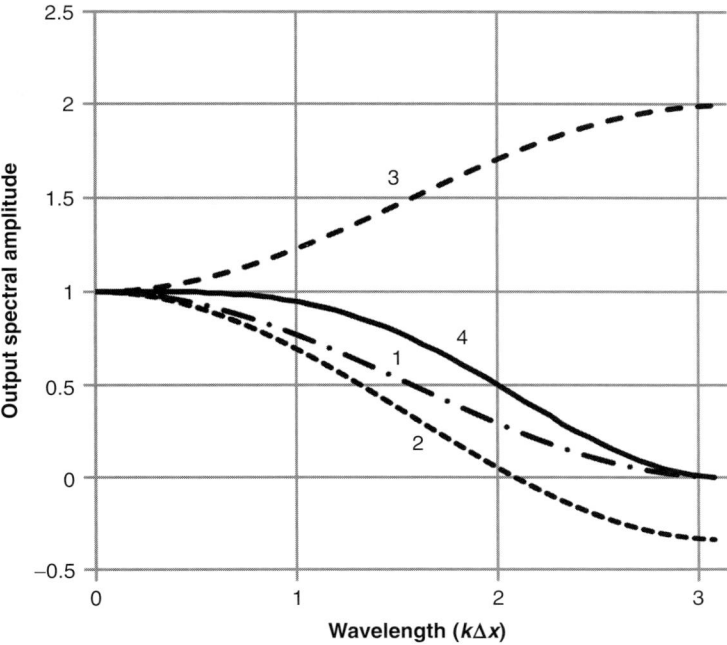

Figure 5.3 Spectral response of a simple smoothing filter.

where i is the imaginary number $\sqrt{-1}$, k is the wavenumber, and A_0 is an arbitrary amplitude. The wavelength is $\lambda = 2\pi/k$. By substituting Eq. (5.11) into the general three-point smoothing operator, the response of the operator can be obtained for any wavenumber. The expression becomes

$$\frac{A_0}{4}[4 + 2se^{-ik\Delta x} - 4s + 2se^{ik\Delta x}], \tag{5.12}$$

which simplifies to

$$A_0[(1 - s) + s\cos k\Delta x]. \tag{5.13}$$

The evaluations of Eq. (5.13) for the three cases described in Eqs. (5.8) through (5.10) are plotted as a function of wavenumber from 0 to π in Figure 5.3. The wavelength at $k\Delta x = \pi$ corresponds to a wavelength of $2\Delta x$, the Nyquist wavelength. The Nyquist wavelength is also referred to as the folding wavelength because all shorter wavelengths alias back to longer wavelengths. The response demonstrates that the weighted average with $s = 0.5$ (1 in Figure 5.3) attenuates a wide range of the larger wavenumber (shorter wavelengths). The simple average of three points with $s = 2/3$ (2 in Figure 5.3) has an adverse effect of changing the sign of the amplitude of the larger wavenumbers (shorter wavelengths), which can lead to a distortion of the shorter wavelengths in the filtered gravity field. The relation showing the optimal removal of folding wavelength noise (4 in Figure 5.3) without changing the smaller wavenumbers (longer wavelengths) of the signal is a sequential

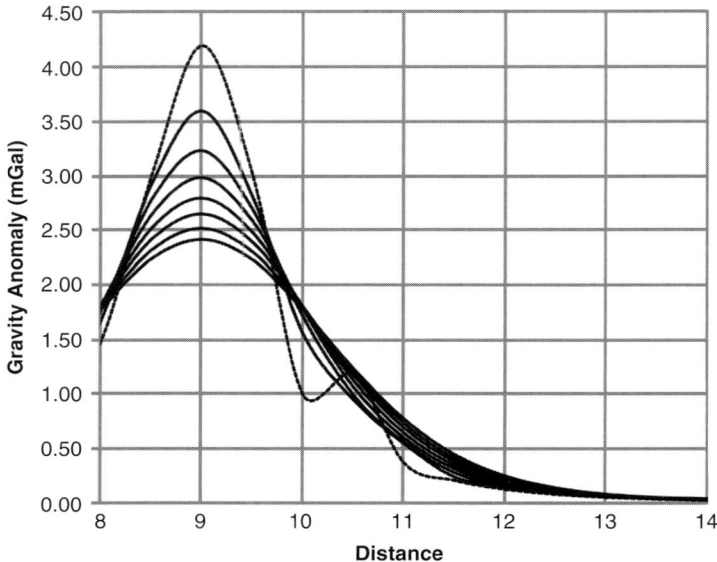

Figure 5.4 Illustration of the filtering of noise superimposed on the anomaly of a sphere. The sphere is at a depth of seven and is a first approximation to a mafic intrusion of density contrast $+0.3$ g/cm^3.

application of smoothing operators with $s = 0.5$ and $s = -0.5$. The wavenumber response for any combination can be obtained by computing the product.

The effects of sequential applications of the filter with $s = 0.5$ and $s = -0.5$ are shown in Figure 5.4. After seven iterations the anomaly has decreased in amplitude by about half, with its signal spread sideways through smoothing. The two anomalous values on the right-hand side, inserted for example as two points in error, are effectively removed after one application of the filter. The data spacing was set at 0.5 distance units so the half-width of the anomaly of a sphere used as an example is approximately two digitizing units. After five iterations of the filter, the filter width exceeds the anomaly width so that the smoothest curve corresponding to the seventh application approaches the width of multiple applications of the filter.

5.5 Orthogonal function decomposition

Orthogonal function decomposition is the separation of an arbitrary function, gravity data in this context, into components that have the mathematical form of individual elements of an orthogonal function set. The advantage of using orthogonal functions is that they provide a simple method of representing the gravity field in terms of functions that are easier to interpret than the raw data or easier to manipulate. The most common application of orthogonal function decomposition is to define and separate regional from local anomalies. Orthogonal functions allow the interpreter to view the data from a different perspective, namely from the perspective of the coordinate system defined by the orthogonal functions. Also, they are common components in the solution to differential equations and in the case of spherical harmonics play an extensive role in expressing the Earth's gravitational field in space.

Orthogonal functions are defined over a domain. The domain may extend from plus to minus infinity or be defined only over a portion of the space. The domain may extend over a rectangular area, the surface of a sphere or an infinite plane and each domain has its own characteristic orthogonal functions. The domain for the Fourier integral transform applied to continuous data extends from plus to minus infinity in one, two, or three dimensions. The domain for the digital equivalent of the Fourier transform, the discrete Fourier transform, extends over a finite number of equally spaced discrete points that in analysis are assumed to repeat continuously to infinity. The domain for the surface spherical harmonics is the surface of a sphere. A set of orthogonal functions is complete if it can define any function within its domain.

By definition, orthogonal functions are linearly independent and provide an alternate coordinate system for observing data. Functions that are linearly independent cannot be expressed as the sum of other members of the set. Given an orthogonal function set with members, $f_1, f_2, f_3, \cdots f_n, f_{n+1} \cdots$, no one of these functions can be expressed as a sum of the others.

For the first member of the set, f_1,

$$f_1 \neq a_2 f_2 + a_3 f_3 + a_4 f_4 + \cdots a_n f_n + a_{n+1} f_{n+1} \cdots, \tag{5.14}$$

indicating that there are no values for the arbitrary constants, a_i that can be found to make the equality hold.

Over their domain of existence, the integral of the product of two orthogonal functions is zero, while the integral of one with itself is a constant. When the functions are defined so that the constant is 1, the functions are orthonormal. For orthonormal functions

$$\int f_i(x) f_j(x) dx = \begin{vmatrix} 0, i \neq j \\ 1, i = j. \end{vmatrix} \tag{5.15}$$

For the purposes of examination and analysis, the gravity field may be expressed as a weighted sum of orthonormal functions

$$\Delta g = \sum_{i=1}^{n} \alpha_i f_i, \tag{5.16}$$

where the equality holds exactly if the orthonormal function set is complete, and where the α_i are found to fit the values of observed gravity. The individual α_i are found by integrating the product of the gravity field and the corresponding orthonormal function in the set. By substituting the expansion of the gravity field into the integral it is clear from Eq. (5.15) that only the ith coefficient remains giving the expression

$$\alpha_i = \int \Delta g(x) f_i(x) dx = \int (\alpha_1 f_1 + \alpha_2 f_2 + \alpha_3 f_3 + \cdots \alpha_n f_n) f_i(x) dx. \quad (5.17)$$

The solutions for the coefficients α_i assume that the gravity anomalies are known completely over the entire domain. If gaps in the data exist, extrapolation methods must be used to fill in missing portions. With gravity data, which are obtained from meter readings at discrete points, there exist only a finite number of discrete values along any profile. Continuous data exist only through extrapolation. When given a discrete number of gravity values, there can only be the same number of independent α_i values, with all others derived from interpolated data. One consequence of discrete sampling is that anomalies at wavelengths shorter or comparable to the sampling interval can be aliased to different, usually longer, wavelengths. Any orthonormal function set should be used with caution where variations in the gravity data or orthonormal functions are not adequately sampled by the given data separation.

The optimal choice of the orthogonal function set depends on the application. There are many to choose from: Chebyshev, Bessel, Laplace, Henkel, to name a few. The domain of the function set would best fit the distribution of the gravity data and the objective of the transformation should be easy to perform in the transform domain. Although orthogonal functions that mimic the character of gravity anomalies could be preferred for interpretation, the choice is most often made on the basis of computational convenience. In particular, the discrete Fourier transform, which is an approximation to the Fourier integral transform, is, perhaps, the most common. The reason for the widespread use of the discrete Fourier transform is its computational efficiency and familiarity with frequency, or equivalently wave number, components of a signal.

5.6 The discrete Fourier transform

The Fourier transform is used to express a continuous field in terms of the amplitudes of individual sinusoidal shapes with different wavenumbers. In Figure 5.5 the individual components are plotted above their combined value. The wavenumber, $k = 2\pi/\lambda$, is usually expressed in terms of the wavelength, λ. The wavelength is the distance between peaks in an anomaly that has the shape of a sine wave. In practice, data are available only for a finite distance and the valid wavenumbers in the Fourier transform must correspond to wavelengths that are even fractions of the length of the gravity data line. Where L is the length of a line of data points, the valid wave numbers are given by $k = 2\pi n/L$, where n is an integer from 0 to ∞. A continuous gravity field, $\Delta g(x)$, can then be equated to a sum of the orthogonal function set consisting of sine and cosine functions of the form

$$\Delta g(x) = a_0 + a_1 \cos \frac{2\pi x}{L} + b_1 \sin \frac{2\pi x}{L} + \cdots + a_n \cos \frac{2\pi nx}{L} + b_n \sin \frac{2\pi nx}{L} \cdots,$$

$$(5.18)$$

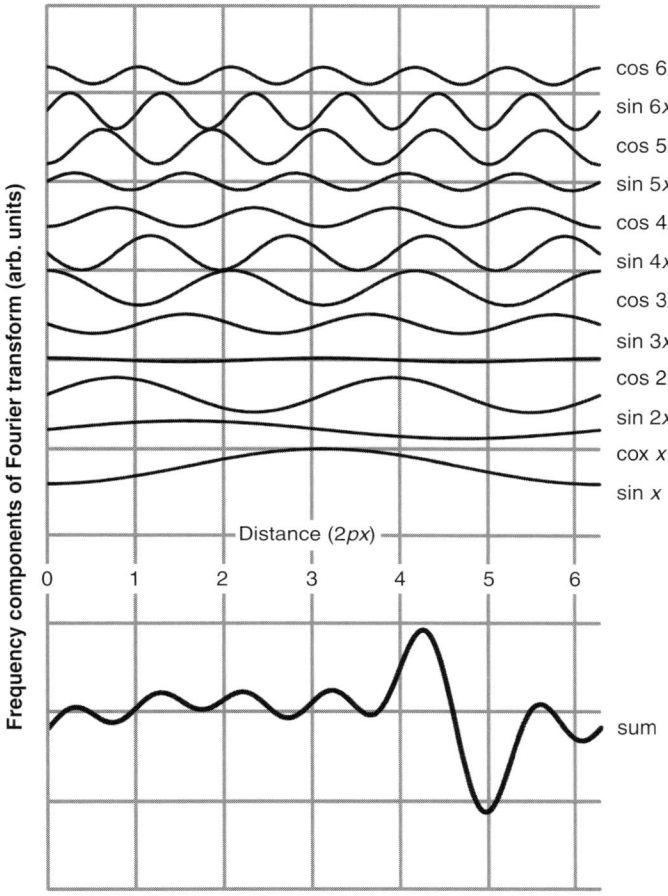

Figure 5.5 An anomalous shape as shown in the bottom trace is the sum of the independent sine and cosine curves shown above.

where for continuous data the n goes to infinity. By considering the coefficients to be complex, the sum can be expressed in the more compact and conventional form as

$$\Delta g\left(x\right) = \sum_{k=1}^{\infty} a_k e^{-i\frac{2\pi kx}{L}}. \tag{5.19}$$

In practice, the sum to infinity is never used.

The sum to infinity implies that the gravity field is continuous and completely defined in the length of data available. Instead, gravity data are obtained as a finite number of discrete points in which the data separation defines, in practice, the shortest wavelengths of data that can be used in this orthogonal function set. The finite number of gravity data values defines the finite number of coefficients that will be unique in the discrete Fourier transform. The coefficients are found by the discrete Fourier transform in the form

$$a_n = \sum_{m=0}^{N-1} \Delta g_m e^{-i\frac{2\pi nm}{N}} \tag{5.20}$$

in which there are N values of gravity at a separation of $\Delta x = L/N$ and, hence, N values of the coefficients, the spectra in the wavenumber domain. For the discrete data the gravity data are expressed as the inverse Fourier transform as

$$\Delta g_m = \Delta g\,(m\Delta x) = \frac{1}{N} \sum_{n=0}^{N=1} a_n e^{i\frac{2\pi nm}{N}}. \tag{5.21}$$

Substitution of Eq. (5.21) for Δg_m into the equation to find the coefficients a_n will show the identity of the discrete Fourier transform pair. In theory N can take any value and most modern computational programs can accommodate arbitrary values of N. However, the fast Fourier transform (FFT) is a computationally efficient way to perform the discrete Fourier transform and many programs use the FFT; however, it works best when N is some power of 2.

While the filtering capabilities of the Fourier transform are unlimited in the wavenumber domain, some applications have the advantage of a simple conceptual application. The most common is the removal of long-wavelength data in order to emphasize shorter, local anomalies. The implied assumption is that the long wavelengths are associated with a regional field. In order to remove a regional field, that can be defined as a constant or long-wavelength anomaly, the gravity values are computed after reducing the amplitude of the first few values of a_n. Normally the values are reduced and not set to zero. Setting portions of the transformed gravity data to zero can lead to oscillation in the field when transformed back to the space domain. Typically a Gaussian-shaped or similar attenuation function is used to minimize oscillation.

Figure 5.6 shows the effect of truncating the higher wavenumbers in the transform domain. In Figure 5.6a the signal is a step function. The simple truncation, the dotted line, introduces considerable oscillation. In contrast, a Gaussian high cut filter applied to the higher wavenumbers retains the general shape of the step function anomaly without generating obvious oscillation. When applied to real data, the effect is the same, but the severity of the distortion depends on the spectra of the data. Figure 5.6b compares the residual signal for data filtered by simple truncation and by a Gaussian filter. Although the effect is not as noticeable, the simple truncation filter (dotted trace) is larger and shows considerable variation in the shape of the shorter wavelengths.

A second application of the Fourier transform in data analysis is the computation of horizontal derivatives of the gravity field. In the wavenumber domain, the first derivative is obtained by multiplying the spectra by ik over a range of $-ik_N$ to $+ik_N$, the Nyquist wavelength. The second derivative is obtained by multiplying twice by ik or equivalently by $-k^2$. By Laplace's equation, the second derivative is the negative of the vertical derivative. The forth derivative, is found by multiplying by k^4 in the wavenumber domain. The fourth derivative is useful in discriminating areas where the gravity field varies rapidly, such as

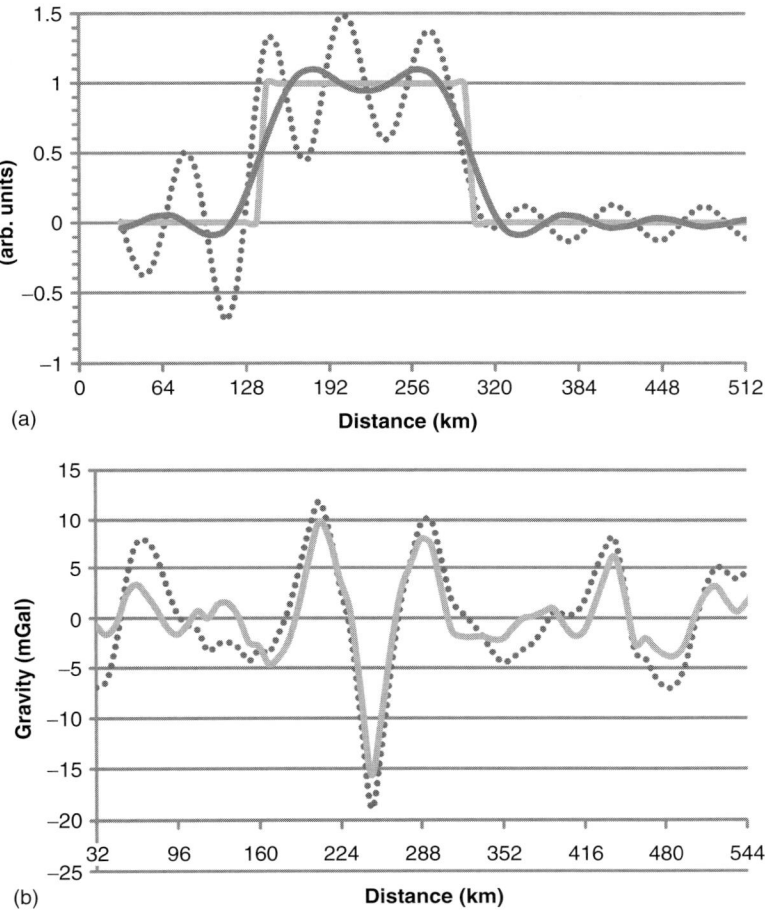

Figure 5.6   Effect of truncation in the wavenumber domain of the Fourier transform. The dotted lines are a simple truncation of shorter wavelengths by setting them to zero. The solid gray line is the original trace. (a) The application to a step function shows the resulting oscillation for truncation. A Gaussian filter gives the dark gray line. (b) With real data the effect is not as noticeable but evident in the shorter wavelengths of data.

near faults or where the anomalous density structure is close to the surface. It will also indicate areas where the noise in the data is abnormally high.

Figure 5.7 shows an example of the application of the fast Fourier transform to compute the first, second and fourth derivatives of an anomaly. For this data, and in most applications, some pre-smoothing is appropriate. These data were first smoothed with the three-point filter described is Section 5.4 with $s = 0.5$. Without some filtering, the second and fourth derivatives can be dominated by the noise at short wavelengths. At 500 km in Figure 5.7 the gravity shows a sharp peak that can be used to illustrate the effects observed on derivative maps. The first derivative, the slope of the line, shows a paired positive and negative peak straddling the anomaly. The peak of the Bouguer anomaly corresponds to the zero crossing

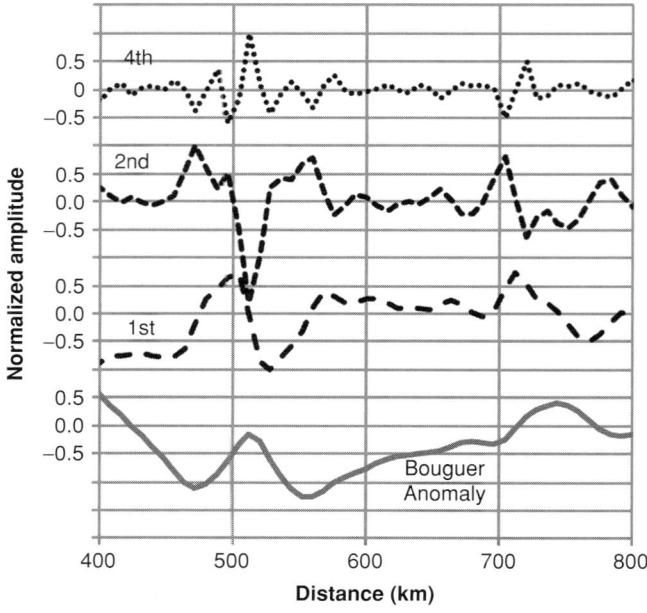

Figure 5.7 Derivatives of a line of gravity data computed in the transform domain using the FFT, The bottom trace is the original data, taken at a separation of 8 km. The amplitudes of the 1st, 2nd, and 4th derivatives and normalized for comparison.

in the first derivative. For a fault-like structure the first derivative would show a single peak with the sign of the peak indicating whether the anomaly increased or decreased going in the positive direction of the axis. The second vertical derivative is symmetric about the anomaly. The second derivative is useful in locating peaks in anomalies from shallow structures.

In two dimensions, the first derivative is a vector with both direction and magnitude. The directionality allows it to be used to enhance anomalies that are aligned in a direction perpendicular to the direction of the derivative. Figure 5.8a is a Bouguer anomaly map of a portion of the South Carolina Coastal Plain. Figure 5.8b is a first-derivative map of Figure 5.8a with the first derivative taken in the northeast direction. In this area most of the structures strike northeast so that the first derivative in the northeast direction will emphasize anomaly trends striking northwest (horizontal in Figure 5.8b). A second-derivative map of the same area is displayed in Figure 5.8c. It shows the effect of a second derivative on local anomalies. The second derivative brings out regions of more rapidly changing anomalies that are likely the edge of the larger geologic units in the crust. It also shows a more irregular field in the area of shallow anomalies. In this example, the original data were smoothed with a three-point filter prior to computing the second derivative in order to suppress the contamination by short-wavelength noise.

In addition to using the direction of the first derivative to suppress a dominant linear trend to the anomalies and enhance anomalies perpendicular to their strike, the Fourier transform can be used to remove a dominant trend by filtering in the wavenumber domain. Figure 5.9a

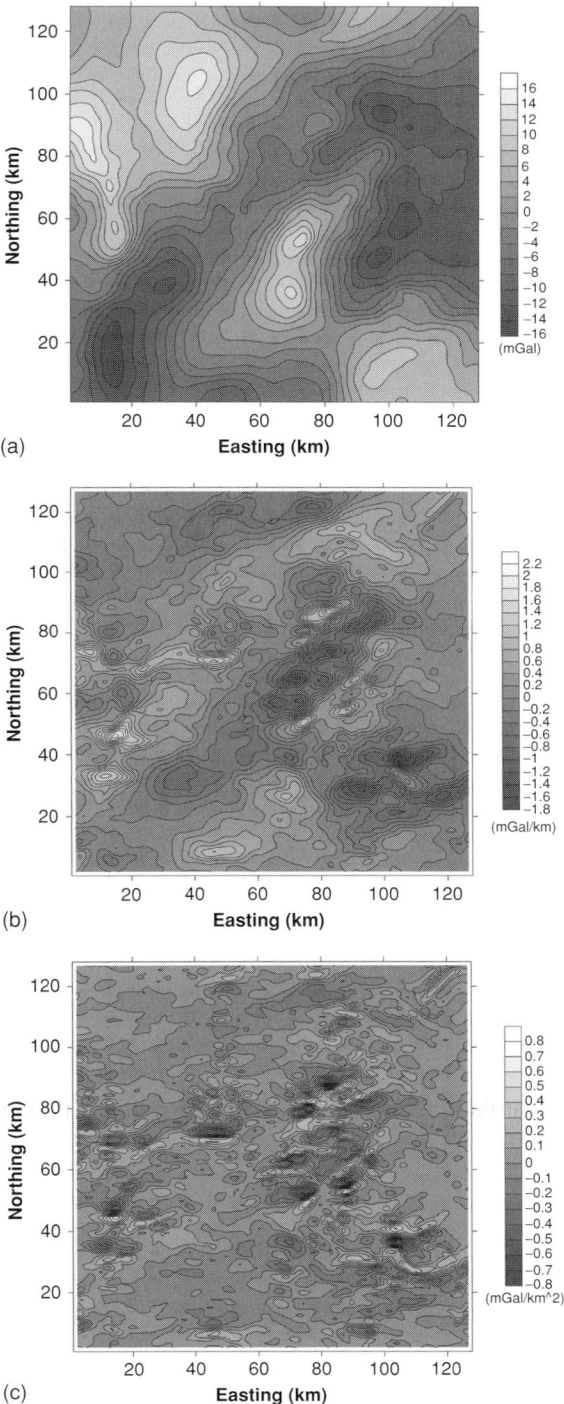

Figure 5.8   (a) Gravity map at a 1 km separation and dimensions of 128 x 128 km. The northing direction is northeast and the easting direction is southeast. (b) First derivative in the northing direction (northeast) of map area shown in Figure 5.8a. (c) Second derivative of the gravity map in Figure 5.8a.

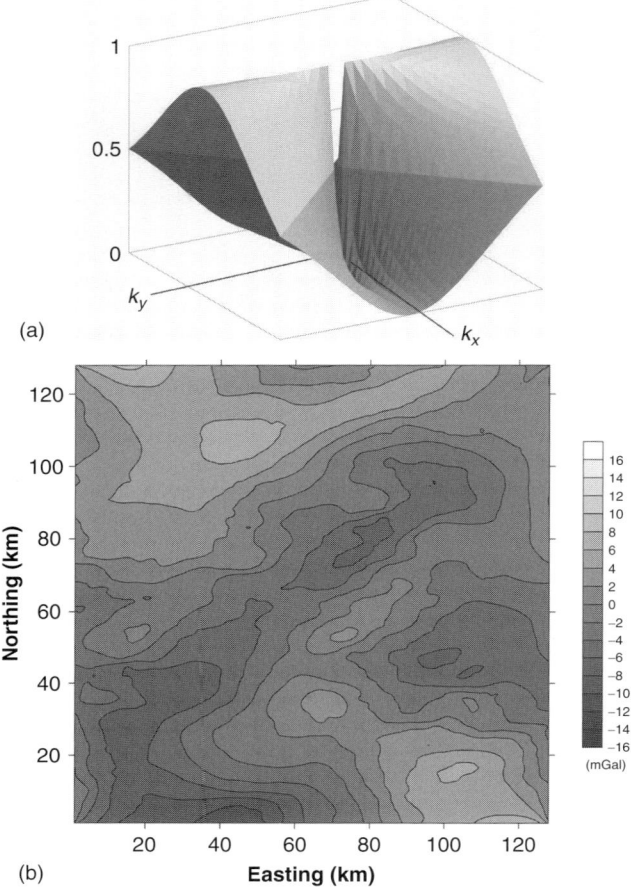

(a)

(b)

Northing (km)

Easting (km)

16
14
12
10
8
6
4
2
0
−2
−4
−6
−8
−10
−12
−14
−16
(mGal)

Figure 5.9   (a) The range of area in the transform domain used to filter out trends. (b) trend removed data from Figure 5.8a .
Figure 5.9b has the same trends attenuated as in Figure 5.8b.

indicates the pie-shaped area of the wave number domain that is attenuated in Figure 5.9b to
suppress wavelength components striking in the northing (northeast) direction. The advantage of wavenumber filtering over the first derivative is that the shorter wavelengths are not
enhanced and the apparent noise is reduced. Filtering in the wavenumber domain is particularly useful to locate structures that are perpendicular to the strike of the major anomalies.

Noise removal is a fourth application of processing data in the Fourier-transform domain.
Noise can be due to random errors in the gravity data acquisition or can be uncorrelated
site specific anomalies related to small structures or topographic variations. Noise is rarely
of interest in the analysis of a survey. The smoothing operator in the time domain is
effective in filtering out the shorter wavelengths. The effect of filtering is to attenuate both
the desired signal and the noise at the short wavelengths. Although this noise is due to
different and perhaps unknown sources, it can often be simulated by white noise in the

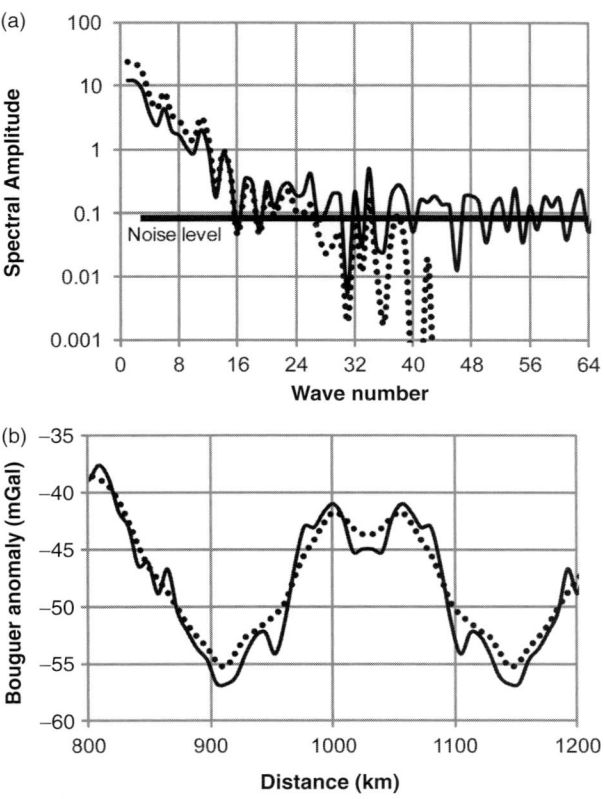

Figure 5.10 The data for this line is a 1024-km sample of observed Bouguer anomalies sampled at an eight km spacing. (a) The spectra of the line showing the effect of short wavelength noise and the spectra after subtraction of the noise (dotted line). (b) A segment of the line comparing the original trace (solid) to the noise-subtracted tract (dotted).

wave number domain. The white noise appears as a base level independent of wavenumber (see Figure 5.10a), which in the transform domain at the shorter wavelengths is often identified as a leveling out of the spectra as the signal from the gravity becomes lower in amplitude than the noise. If a base or noise level can be identified, it may be subtracted from the total signal to give a noise-reduced signal Figure 5.10b. This "noise" is removed by computing the gravity values from wave numbers that correspond to wavelengths that are longer than the noise or from spectral amplitudes that are reduced by an amount equivalent to the noise level. The application shown in Figure 5.10 involved first identifying the noise level. The attenuation filter in the wavenumber domain is defined by first computing an average or smoothed spectra of the gravity data and secondly computing the ratio of the smoothed spectra minus the noise level to the smoothed spectra. The resulting output is a smoother trace. The amount of smoothing will depend on the noise level chosen and the degree of smoothing of the Bouguer anomalies, parameters that are determined by trial and error.

5.7 Least squares criteria for fitting data

In Section 5.5, orthogonality was used as a criterion for fitting orthogonal functions to a gravity field. When a complete orthogonal function set is not used, the fit of a few orthogonal functions can be optimized by applying the criteria that the fit minimizes the square error. An appropriate set of linearly independent functions is needed for fitting to the data. The coefficients obtained are not the same as those obtained for orthogonal function decomposition. Also, they differ when different numbers of orthogonal functions are used. Orthogonal function decomposition gives components that are independent of the number of functions. For two-dimensional data the approximation to the gravity field can be expressed by the equations

$$\Delta g\left(x, y\right) = \sum_{i=1}^{n} k_i f_i\left(x, y\right),\tag{5.22}$$

where the f_i are linearly independent functions. To solve for the coefficients k_i we set up the equations in matrix form. This formulation assumes that the gravity data are equally spaced. The equations would need to be modified with weights for the area represented by the gravity data or the gravity data would need to be extrapolated to equally spaced points. For m observations of gravity, each observation can be expressed as the sum of the products of the coefficients and its orthogonal function, as

$$\Delta g_j = k_1 f_1\left(x_j, y_j\right) + k_2 f_2\left(x_j, y_j\right) + \cdots + k_n f_n\left(x_j, y_j\right)\tag{5.23}$$

$$\begin{vmatrix}\Delta g_1\\ \vdots\\ \Delta g_m\end{vmatrix} = \begin{vmatrix}\begin{pmatrix}f_1\left(x_1, y_1\right) & \cdots & f_n\left(x_1, y_1\right)\\ \vdots & \ddots & \vdots\\ f_1\left(x_m, y_m\right) & \cdots & f_n\left(x_m, y_m\right)\end{pmatrix}\end{vmatrix}\begin{vmatrix}k_1\\ \vdots\\ k_n\end{vmatrix}.\tag{5.24}$$

The number of equations, m, should significantly exceed the number, n, of orthogonal functions used in order to assure that the inversion for the coefficients is stable. In two-dimensional applications, the stability of the equations can fail if, for example, the data are constrained to a one-dimensional line. The simple solution for overdetermined conditions for the coefficients, k, can be derived in matrix formulation first by expressing Eq. (5.24) as

$$\Delta g = FK\tag{5.25}$$

and then pre-multiplying by the transpose of F and then by the inverse of $F^T F$ to get

$$\left(F^T F\right)^{-1} F^T \Delta g = \left(F^T F\right)^{-1}\left(F^T F\right) K = K.\tag{5.26}$$

Other inversion equations, those designed to identify singular conditions, can also be used, but for most applications of this type, the solution is for a limited number of coefficients and the solution should be stable.

5.8 Polynomials

Polynomial functions are common in applications of least square fitting to orthogonal function sets. The two-dimensional power series in the coordinates x and y can be written as

$$\Delta g\,(x,\,y) = \sum_{i=0}^{m} \sum_{j=0}^{n} a_{ij} x^i y^j. \tag{5.27}$$

In the case where $m = n = 2$, Eq. (5.27) is truncated to

$$\Delta g\,(x,\,y) = a_{00} + a_{01} y + a_{10} x + a_{02} y^2 + a_{11} xy + a_{20} x^2. \tag{5.28}$$

A common problem with polynomial fits to data is that the edges do not behave well. Their application should be limited to data that are constrained over the entire map area and have characteristics similar to the polynomials. For example, as the coordinates approach infinity, the anomalies should also get very large to match the increase in amplitude with higher powers of the polynomial.

The edge effect is unavoidable in most applications, not just with the polynomial fits to data. When applying the smoothing operator, information is typically missing outside the map area, leading to distortion and a change in the character of the smoothing along the edges. The Fourier transform is cyclic so that features on one side of the map will be carried over to the opposite side. A partial solution to edge effects is to extend the data to points outside the study area so that discontinuities at the edge are minimized. Although this does not completely remove the edge effect, extension into a larger area with a taper to a mean value will limit the influence of edge effects on interpretations of data in the study area.

5.9 Upward continuation

The gravity field for the Earth is continuous throughout all free space above the Earth's surface; however, measurements of the Earth's gravity field are generally restricted to its surface. The process of finding the Earth's gravity field, or any potential field, above the surface of measurement (that is further from the mass that generates the field) is referred to as upward continuation. Because upward continuation finds the field at a greater distance from the source of the field, upward continuation is a smoothing operation that transforms the anomalous fields of local anomalies to broader anomalies. These broader anomalies may be used as one definition of regional anomalies. Upward continuation has the advantage of preserving the mathematical properties of a potential field. In this section the known gravity field is assumed to exist on a flat surface. Gravity data reductions use the elevation and Bouguer plate correction to move the gravity values at the point of observation to the geoid or ellipsoid surface without concern for changes in the potential field with elevation. The changes in the anomalous field with elevation for most areas are slight, principally

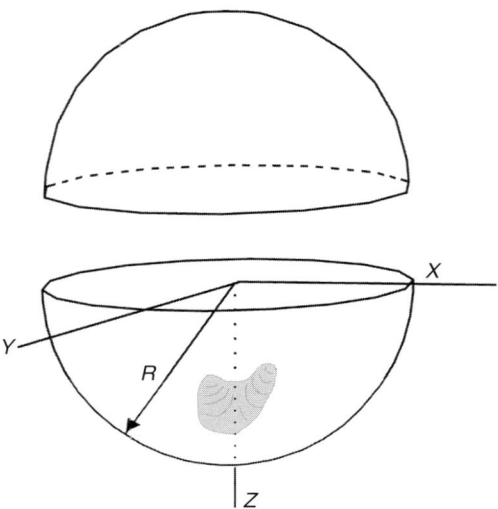

Figure 5.11   Application of Green's third identity to a hemisphere.

because the changes in elevation are small compared to the data separation. Hence, in this section it will be assumed that the field to be continued is on a flat surface. This condition does not hold in areas of extreme topography where alternate techniques, such as analytical continuation, are needed.

The derivation of the equations for upward continuation starts with Green's third identity. Green's third identity provides a relation between the potential field in space and the potential field on a surface. The analysis here is appropriate for local surveys, those where the curvature of the Earth is negligible. The surface for analysis is the surface on which the data are originally known, which is assumed to be a plane. In order to compute the volume for integration, a hemisphere as shown in Figure 5.11 is used to surround the attracting mass and its radius, as well as the dimensions of the flat surface, are extended to infinity. Green's third identity can be written in the form

$$\iiint \left(U \nabla^2 \frac{1}{l} - \frac{1}{l} \nabla^2 U \right) dv = \iint \left(U \frac{\partial}{\partial n} \frac{1}{l} - \frac{1}{l} \frac{\partial U}{\partial n} \right) ds. \tag{5.29}$$

The surface of interest is the flat part of the hemisphere, or above. Hence, the distance, l, from the mass inside the hemisphere to these surfaces is greater than zero. The first term in the volume integral will be identically zero because $1/l$ can be shown to be harmonic in the region where $l \neq 0$ and, hence

$$\nabla^2 \frac{1}{l} = 0. \tag{5.30}$$

Green's third identity reduces to the form

$$\iiint \frac{1}{l} \nabla^2 U \, dv = \iint \left(U \frac{\partial}{\partial n} \frac{1}{l} - \frac{1}{l} \frac{\partial U}{\partial n} \right) ds. \tag{5.31}$$

In the interior of the mass, Poisson's differential equation has a non-zero value given by

$$\nabla^2 U = 4\pi G\rho. \tag{5.32}$$

On substitution of Poisson's differential Eq. (5.32) into Eq. (5.31), the integral on the left has the same form as the integral for the potential

$$U(x, y, z) = G \iiint \frac{\rho}{l} dv, \tag{5.33}$$

except for the factor of 4π. Green's third identity then takes the form

$$-\iiint \frac{4\pi G\rho}{l} dv = -4\pi U = \iint \left(U \frac{\partial}{\partial n} \frac{1}{l} - \frac{1}{l} \frac{\partial U}{\partial n} \right) d\sigma. \tag{5.34}$$

The right-hand side of Eq. (5.34), the integral over the surface, can be evaluated by separating the integration over the flat surface from the integration over the hemisphere. The integration over the flat surface is then extended to infinity. The integral over the hemisphere is evaluated by taking the limit as the radius goes to infinity. In the limit as R goes to infinity

$$\lim_{R \to \infty} \left(\frac{1}{l} \right) = \frac{1}{R} \tag{5.35}$$

and

$$U \to \frac{GM}{R}, \quad \frac{\partial}{\partial n} \left(\frac{1}{l} \right) \to \frac{\partial}{\partial R} \left(\frac{1}{R} \right) \to \frac{1}{R^2}, \tag{5.36}$$

where the normal is in the direction of the radius R and where M is the total mass of the anomaly. At infinity, M approximates a point mass. These relations allow the components in the integral over the hemisphere to be written as

$$-4\pi U = \int\int_{-\infty}^{\infty} \left(U \frac{\partial}{\partial z} \frac{1}{l} - \frac{1}{l} \frac{\partial U}{\partial z} \right) d\sigma + \lim_{R \to \infty} \iint \left(\frac{M}{R} \frac{1}{R^2} - \frac{1}{R} \frac{M}{R^2} \right) d\sigma. \tag{5.37}$$

At infinity, then, the surface integration over the hemisphere goes to zero, leaving the integration over the flat surface that corresponds to the observed gravity field. On the flat surface, the normal direction corresponds to the z-, or vertical, axis, giving

$$-4\pi U = \iint \left(U \frac{\partial}{\partial z} \frac{1}{l} - \frac{1}{l} \frac{\partial U}{\partial z} \right) ds. \tag{5.38}$$

A similar expression can be obtained for an upper hemisphere that contains no mass. The lack of mass in the upper hemisphere requires that the term $-4\pi U$ equal zero. The change in sign relates to the change in the direction of the normal across the flat surface. The resulting expression is

$$0 = \iint \left(U \frac{\partial}{\partial z} \frac{1}{l} + \frac{1}{l} \frac{\partial U}{\partial z} \right) d\sigma, \tag{5.39}$$

and it may be subtracted from Eq. (5.38) for the lower hemisphere to eliminate the first term and provide an integral expression for the potential on any surface above the flat surface in the form

$$U(x, y, z) = \frac{1}{2\pi} \iint \left(\frac{1}{l} \frac{\partial U}{\partial z} \right) d\sigma. \tag{5.40}$$

Equation (5.40) is in terms of the vertical gradient of the potential on the surface. The vertical gradient of the potential in Eq. (5.40) will give the expression for the gravity anomaly and the distance $l = ((x^2 - \xi^2) + (y^2 - \eta^2) + z^2)^{1/2}$ is from the point of computation above the surface (x, y, z) to the gravity anomaly in the integral and on the surface $(\xi, \eta, 0)$

$$\Delta g = \frac{\partial}{\partial z} U = \frac{1}{2\pi} \iint \Delta g_s \frac{\partial}{\partial z} \left(\frac{1}{l} \right) d\sigma \tag{5.41}$$

or

$$\Delta g(x, y, z) = \frac{z}{2\pi} \iint \frac{\Delta g(\xi, \eta, 0)}{l^3} d\sigma, \tag{5.42}$$

where Δg_s is the gravity observed on the surface and where

$$\frac{\partial}{\partial z} \left(\frac{1}{l} \right) = \frac{-z}{l^3}. \tag{5.43}$$

The evaluation of the continuation integral gives the gravity anomaly at a distance of z above the surface where Δg is known. In Eq. (5.42) the denominator approaches zero as z approaches zero. In the case where the gravity anomaly is a constant independent of position, the integral approaches $\frac{z}{2\pi} \Delta g \frac{2\pi}{z}$ showing that a constant gravity anomaly on a flat plane has a zero vertical gradient and is continued upward without change. Consequently, by adding and subtracting the value of the gravity anomaly at the point of computation, Eq. (5.42) can be written in the form

$$\Delta g(x, y, z) = \Delta g(x, y, 0) + \frac{z}{2\pi} \iint \frac{\Delta g_s(\xi, \eta, 0) - \Delta g(x, y, 0)}{l^3} d\sigma'. \tag{5.44}$$

Equation (5.44) is appropriate for a flat plane. In the case of the upward continuation of gravity anomalies on a spherical surface, such as the Earth, the expression equivalent to Eq. (5.42) from Heiskanen and Moritz (1967) has the form

$$\Delta g = \frac{R^2}{2\pi r} \iint \left[\frac{(r^2 - R^2)}{l^3} - \frac{1}{r} - \frac{3 \cos \phi}{r^2} \right] \Delta g_s d\sigma, \tag{5.45}$$

where R is the radius to the surface of the sphere, r is the radial distance to the computational point, and ϕ is the angle between r and R. The first term is the continuation integral for a potential function and the second two terms account for the removal of theoretical gravity in Δg. This is the appropriate form for continuing the gravity field upward at distances where the curvature of the Earth becomes significant.

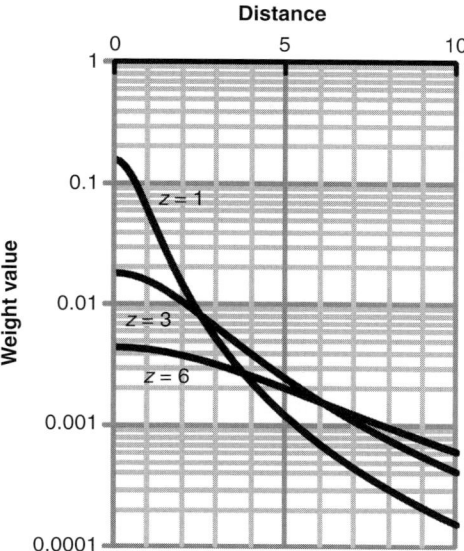

Figure 5.12 Shape of the weight function for upward continuation.

5.10 Numerical integration of upward continuation

The gravity field is rarely simple enough to express as a mathematical expression. In many applications, upward continuation reverts to numerical evaluation of the continuation integral in Eq. (5.44). Upward continuation may be considered a special case of a weighted average of the gravity field on the surface where the weight function is

$$\frac{z}{2\pi l^3}. \tag{5.46}$$

When z is small and at positions where r is small the weight is proportional to $1/l^3$, which approaches a singularity, giving large values relative to more distant values. Figure 5.12 demonstrates the shape of $\frac{z}{2\pi l^3}$ at a selection of heights. In continuation to slight heights, the anomalies are changed only slightly. For slight heights, the field is constant, with little or no curvature, and would not be expected to change significantly with elevation. In that case Eq. (5.44) can be written as

$$\Delta g\,(x,\,y,\,z) = \Delta g(x,\,y,\,0) + z\frac{\partial}{\partial z}\left(\Delta g_s(\xi,\,\eta,\,0) - \Delta g(x,\,y,\,0)\right), \tag{5.47}$$

where it is seen that the integral is equivalent to the vertical derivative that gives the correct upward continuation of the gravity field. For z small, or zero, the integral gives the vertical derivative on the surface. When z is large, a much wider area is included in the average yielding a smoother version of the gravity field. For numerical integration of Eq. (5.44) the gravity field should be defined with the same spatial definition (or grid increment) as is required to accurately represent the function $z/2\pi l^3$. Upward continuation then is best computed for heights greater than the sampling interval of the gravity data.

Evaluation of the integral when z is small can be approximated by defining the gravity field surrounding each grid point with a Taylor's expansion

$$\Delta g_s - \Delta g(x, y, 0) = (\Delta g_{nm} - \Delta g(x, y, 0)) + (\xi - x)\frac{\partial}{\partial x}\Delta g_{nm} + (\eta - y)\frac{\partial}{\partial y}\Delta g_{nm}$$
$$+ \frac{1}{2}\left[(\xi - x)^2\frac{\partial^2}{\partial x^2}\Delta g_{nm} + (\eta - y)^2\frac{\partial^2}{\partial y^2}\Delta g_{nm} + 2(\xi - x)(\eta - y)\frac{\partial^2}{\partial y\partial x}\Delta g_{nm}\right],$$

$$(5.48)$$

where $m\Delta x$ and $n\Delta y$ define the position of the grid point under evaluation and $\Delta g_{nm} = \Delta g(\xi_m, \eta_n, 0)$ is the gravity or its derivatives evaluated at the grid point. The derivatives can be evaluated using finite-difference approximations and are constants that can be taken out of the integral leaving integrals that can be evaluated in closed form. The upward continuation is then converted to a summation. For large values of z, the usual case in interpretation of gravity anomalies, only the first term is needed.

5.11 Fourier transform continuation

The Fourier integral transform can be used to formulate a continuation scheme that is useful for both upward and downward continuation. Upward and downward continuation by the discrete Fourier transform is limited by the size of the grid and the cyclic properties of the discrete Fourier transform. In downward continuation the value of gravity is known at some elevation z and the object is to find the value of the gravity field at a position closer to the attracting masses. In this case, Eq. (5.42) becomes an integral equation that is solved by transforming it into the wavenumber domain. First, the two-dimensional Fourier transform of Eq. (5.42) is written as

$$\mathcal{F}\{\Delta g(x, y, z)\} = \frac{z}{2\pi}\iint \Delta g(\xi, \eta, 0)\iint \frac{1}{l^3}e^{-i(px+qy)}d\xi d\zeta dx dy,\qquad (5.49)$$

where p and q are the wavenumbers in the x and y directions.

In Eq. (5.49), if we let

$$x = \xi + r\cos\theta \qquad\qquad (5.50)$$
$$y = \eta + r\sin\theta \qquad\qquad (5.51)$$
$$p = u\cos\phi \qquad\qquad (5.52)$$
$$q = u\sin\phi. \qquad\qquad (5.53)$$

Then Eq. (5.49) can be recognized as

$$\mathcal{F}\{\Delta g(x, y, z)\} = \frac{z}{2\pi}\iint \Delta g(\xi, \eta, 0)\iint \frac{e^{-i(p\xi+q\eta+ru\cos(\theta-\phi))}}{\left((r\cos\theta)^2 + (r\sin\theta)^2 + z^2\right)^{3/2}}d\xi d\zeta r dr d\theta$$

$$(5.54)$$

$$\mathcal{F}\{\Delta g(x, y, z)\} = \frac{z}{2\pi} \iint \Delta g\,(\xi, \eta, 0)\,e^{-i(p\xi+q\eta)}d\xi d\zeta \int\limits_{r=0,\theta=0}^{\infty,2\pi} \frac{e^{-i(ru\cos(\phi-\theta))}}{(r^2 + z^2)^{\frac{3}{2}}} r dr d\theta$$

$$(5.55)$$

$$\mathcal{F}\{\Delta g(x, y, z)\} = \frac{z}{2\pi} \iint \Delta g\,(\xi, \eta, 0)\,e^{-i(p\xi+q\eta)}d\xi d\zeta \int\limits_{0}^{\infty} \frac{J_0\,(ur)}{(r^2 + z^2)^{\frac{3}{2}}} r dr \qquad (5.56)$$

$$\mathcal{F}\{\Delta g(x, y, z)\} = \frac{z}{2\pi} \iint \Delta g\,(\xi, \eta, 0)\,e^{-i(p\xi+q\eta)}d\xi d\zeta \frac{2\pi}{z}e^{-uz}, \qquad (5.57)$$

and since $u = \sqrt{p^2 + q^2}$

$$\mathcal{F}\{\Delta g(x, y, z)\} = e^{-(p^2+q^2)^{1/2}z} \iint \Delta g\,(\xi, \eta, 0)\,e^{-i(p\xi+q\eta)}d\xi d\zeta \qquad (5.58)$$

$$\mathcal{F}\{\Delta g(x, y, z)\} = e^{-(p^2+q^2)^{1/2}z}\mathcal{F}\{\Delta g(x, y, 0)\}. \qquad (5.59)$$

A notable characteristic of Eq. (5.59) is that in downward continuation, where z is negative, the shorter wavelengths are amplified in proportion to the product of their wavenumber and the depth of continuation. Any sources of anomalous field in the depth range of downward continuation will be highly amplified. Hence, downward continuation does not work well when the gravity field has components with sources at shallow depth. This also holds for noise, which replicates sources at all depths. For this reason data used in downward continuation should be filtered to minimize the influence of noise and shallow sources.

5.12 Finite-difference methods

When gravity data are available at points distributed equally over an area with values defined on a regular grid, the finite-difference technique may be used in continuation. The restriction is, again, that the field cannot be continued through any mass or short wavelength noise that contributes to the field. We use the fact that in free space, the Laplacian of a potential function is zero

$$\nabla^2 \Delta g\,(x, y, z) = 0 = \frac{\partial^2}{\partial x^2}\Delta g\,(x, y, z) + \frac{\partial^2}{\partial y^2}\Delta g\,(x, y, z) + \frac{\partial^2}{\partial z^2}\Delta g\,(x, y, z)\,z. \quad (5.60)$$

To Eq. (5.60) we apply the first-order finite-difference approximation to the second derivative. For the z component after dropping higher-order terms the finite-difference approximatio is

$$\frac{\partial^2}{\partial z^2}\Delta g \cong \frac{\Delta g\,(x, y, z - \Delta z) - 2\Delta g\,(x, y, z) + \Delta g\,(x, y, z + \Delta z)}{\Delta z^2}. \qquad (5.61)$$

In general, the spacing in the x and y directions are set equal. While this is not necessary, it does simplify the stability criteria for downward continuation. Equation (5.61) when inserted into Eq. (5.60) allows the gravity anomaly at one elevation to be expressed in terms of the gravity anomaly at two adjacent levels. Hence, given the gravity anomaly at two elevations, the field may be continued up or down one increment in elevation. Starting with the field at zero elevation, the usual procedure is to use Eq. (5.42), Eq. (5.44) or the Fourier transform method to continue the field upward, and then use the continued field in the finite-difference method to continue the field downward. Hence, given the field continued up, one can use finite differences to compute the field continued down. Similarly, given the field continued down, one can use finite differences to compute the field continued up. Substituting Eq. (5.61) and similar expressions for the x and y components the gravity field continued downward to position $-\Delta z$ can be expressed in terms of the gravity field at the current level and at a position above the line at $+\Delta z$, for a point at the origin as

$$
\begin{aligned}
\Delta g\,(0, 0, -\Delta z) = {}& 2\Delta g\,(0, 0, 0) - \Delta g\,(0, 0, +\Delta z) \\
& + \frac{\Delta z^2}{\Delta x^2}\,(2\Delta g\,(0, 0, 0) - \Delta g\,(-\Delta x, 0, 0) - \Delta g\,(+\Delta x, 0, 0)) \\
& + \frac{\Delta z^2}{\Delta y^2}\,(2\Delta g\,(0, 0, 0) - \Delta g\,(0, -\Delta x, 0) - \Delta g\,(0, +\Delta x, 0))\,. \quad (5.62)
\end{aligned}
$$

Evaluation of the downward continued field requires knowledge of the field above the surface in free space. In viewing downward continued data, the field is successively continued down to greater depths until the field shows instability. Figure 5.13 shows the finite-difference technique applied to a horizontal cylinder where the two levels were generated using a theoretical model. As the downward continuation approaches the center of the sphere, the continued field becomes unstable.

5.13 Analytical continuation

In all the continuation methods discussed above, the gravity anomalies were assumed to be defined on a plane of constant elevation. That is not always a correct assumption. In areas of more extreme topography, the anomalies observed at the higher elevations would have a shape referenced to the elevation of the observations, not the flat surface of the ellipsoid or geoid. The gravity reduction brings the anomaly down to the ellipsoid or geoid but does not apply a downward continuation to the anomalous part of the field. In effect, the standard free air gravity reduction moves the masses with shallow depths to positions below the ellipsoid or geoid. The significance is that by assuming the anomalies are observed on the ellipsoid or geoid, the shape of the Earth's field continued to above the surface will differ from the actual field. For local surveys in areas of modest topography, the difference is not significant. Also, from the point of view of the interpreter of these structures, referencing these anomalies to the physical surface is appropriate. However, the difference can be important in areas of extreme topography and when matching ground data with data at the elevation of satellites. Analytical continuation computes the gravity anomaly on a flat surface from anomalies distributed on a surface with topographic relief. The field on a flat surface determined by analytic continuation is the field that gives the correct field above

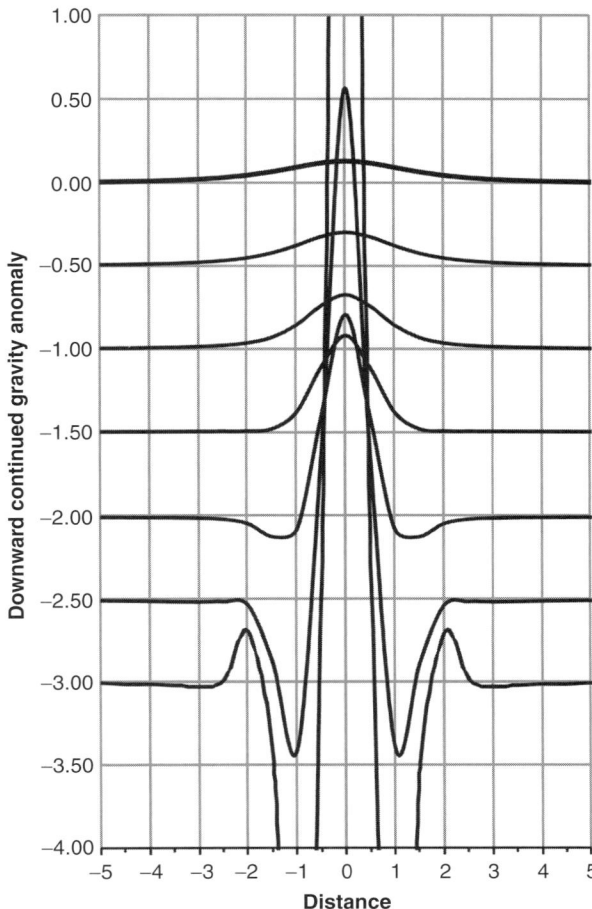

Figure 5.13 Downward continuation using the finite difference technique. The starting function was generated using the gravity anomaly for a sphere at two heights.

the surface. In effect, the analytically continued field is the field that when continued up to the surface gives the observed anomaly and hence the correct anomaly for all space. The analytically continued field is the appropriate anomaly field to use in the Stokes integral for computation of the geoid and deflections of the vertical on a level surface as opposed to the physical surface. Also, it is the appropriate reduction to use when modeling structures in mountainous terrain when the gravity observations are on the physical surface.

Analytical continuation is an approximate solution to the integral equation

$$\Delta g\left(x, y, z(x, y)\right) = \Delta g_0(x, y, 0) + \frac{z}{2\pi} \iint \frac{\Delta g_0(\xi, \eta, 0) - \Delta g_0(x, y, 0)}{l^3} d\sigma, \quad (5.63)$$

where $\Delta g(x, y, z(x, y))$ is the known function as measured on the physical surface, at elevation $z(x, y)$ and $\Delta g_0(x, y, 0)$ is the gravity anomaly on the flat surface as shown in

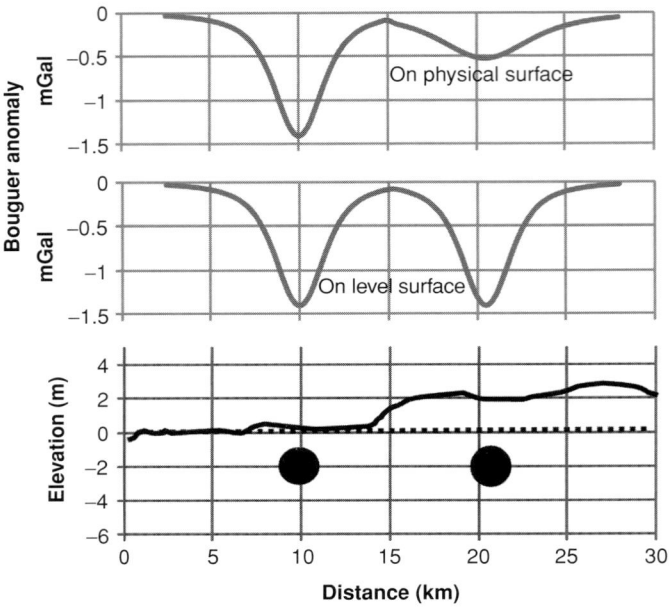

Relative positions of the observed gravity field and the downward continued field using analytical continuation. The possibility of anomalous mass above the downward continued field always leads to a displacement of the anomalous mass or instabilities.

Figure 5.14. The significance of $\Delta g_0(x, y, 0)$ is that it is the function that when continued upward, gives the anomalies observed at the surface.

The approximate solution to Eq. (5.63) is obtained by expanding the gravity anomalies in a Taylor series

$$\Delta g\left(x, y, z\right) = \Delta g\left(x, y, 0\right) + \left[\sum_{n=1}^{\infty} \frac{1}{n!} z^n \frac{\partial^n}{\partial z^n}\right] \Delta g\left(x, y, 0\right). \tag{5.64}$$

The principal difference between Eq. (5.64) and Eq. (5.63) is that the derivatives are at the surface, whereas in Eq. (5.63) the integral is the equivalent derivative from 0 to z that gives the gravity field at elevation $z(x, y)$. Using Eq. (5.64) we can define a linear operator, U, for upward continuation

$$\Delta g\left(x, y, z\right) = \left[I + \sum_{n=1}^{\infty} \frac{1}{n!} z^n \frac{\partial^n}{\partial z^n}\right] \Delta g\left(x, y, 0\right) = U \Delta g\left(x, y, 0\right), \tag{5.65}$$

where U takes the form

$$U = I + \sum_{n=1}^{\infty} z^n L_n \tag{5.66}$$

and where

$$L_n = \frac{1}{n!} \frac{\partial^n}{\partial z^n}. \tag{5.67}$$

In order to solve for the anomaly on the surface from the anomaly at elevation z, we need the downward continuation operator, $D = U^{-1}$. The technique for finding the downward continuation operator is to replace z by kz so that

$$U = I + \sum_{n=1}^{\infty} k^n z^n L_n = \sum_{n=0}^{\infty} k^n U_n. \tag{5.68}$$

Then $U_0 = I$, $U_1 = zL_1$, and in general, $U_n = z^n L_n$. A similar notation can be used for the downward operator and by using the identity

$$I = UD = \sum_{p=0}^{\infty} k^p U_p \sum_{q=0}^{\infty} k^q D_q = \sum_{p=0}^{\infty} \sum_{q=0}^{\infty} k^{p+q} U_p D_q. \tag{5.69}$$

The identity must hold for all powers of k independently, so all the coefficients of k are identically equal to zero, giving for $n = 0$ and for all n

$$U_0 = D_0 = I$$
$$D_n + \sum_{r=1}^{n} U_r D_{n-r} = 0, \tag{5.70}$$

and

$$D_0 = I$$
$$D_1 = -U_1 D_0 = -U_1$$
$$D_2 = -U_1 D_1 - U_2 D_0 = -U_1 D_1 - U_2 \tag{5.71}$$
$$D_n = -\sum_{r=1}^{n} U_r D_{n-r}.$$

Now, applying Eq. (5.71) to Eq. (5.65) the solution for gravity at $z = 0$ is

$$\Delta g(x, y, 0) = DU \Delta g(x, y, 0) = D\Delta g(x, y, z)$$
$$\Delta g(x, y, 0) = \sum_{n=0}^{\infty} D_n \Delta g(x, y, z) \tag{5.72}$$
$$\Delta g(x, y, 0) = \Delta g(x, y, z) - zU_1 \Delta g(x, y, z) + U_1 U_1 \Delta g(x, y, z) - U_2 \Delta g(x, y, z) + \cdots.$$

Then, using the definitions of U_n and L_n in Eq. (5.66) and Eq. (5.67)

$$\Delta g(x, y, 0) = \Delta g(x, y, z) - z\frac{\partial}{\partial z}\Delta g(x, y, z) + \frac{z^2}{2}\frac{\partial}{\partial z}\left(\frac{\partial}{\partial z}\Delta g(x, y, z)\right) + \cdots. \tag{5.73}$$

In order to evaluate Eq. (5.73) an expression for the vertical derivative is needed. The integral form of the vertical derivative can be used, by adapting the integral portion of Eq. (5.63), which evaluated at the elevation of the observed gravity anomaly takes the form

$$\frac{\partial}{\partial z}\Delta g(x, y, z) = \frac{1}{2\pi} \iint \frac{\Delta g(\xi, \eta, z) - \Delta g(x, y, z)}{((x - \xi)^2 + (y - \eta)^2)^{3/2}} d\sigma. \tag{5.74}$$

Equation (5.74) can be evaluated as outlined in the derivation for Eq. (5.48). In this case the difference $\Delta g(\xi, \eta, z) - \Delta g(x, y, z)$ is expanded into a power series in which the integrals over small rectangular elements can be evaluated in closed form. In Eq. (5.73) the second derivative can be evaluated directly in finite-difference form from the relation

$$0 = \frac{\partial^2}{\partial x^2}\Delta g + \frac{\partial^2}{\partial y^2}\Delta g + \frac{\partial^2}{\partial z^2}\Delta g \tag{5.75}$$

as evaluated in Eq. (5.62).

Whether using analytical continuation, or conventional downward continuation there is still a skeleton in the closet. The possibility that the sources of the gravity anomaly are above the surface $\Delta g(x, y, 0)$ always poses a potential problem with stability. Any density anomaly that has the appearance of a point source will cause instability in the downward continuation. Consequently, the downward continued surface cannot always be used as a valid representation of the anomaly from these shallow structures.

6 Interpretation of density structure

6.1 Introduction

The density distribution required to generate a gravity anomaly is non-unique. While a density distribution generates a unique anomaly, there are many density distributions that can generate that same anomaly. Hence, gravity anomalies alone cannot determine the distribution of densities. However, because the gravity field is generated by a distribution of single poles of attraction, it is generally simpler to interpret than dipole fields like the magnetic field. The non-uniqueness of the potential field prevents one from obtaining an unconstrained solution for the density structure. The only parameter that can be defined from noise-free data is the maximum depth to some part of the structure, and that useful information is itself based on the assumption that part of the anomaly is equivalent to a point mass. Also, if the gravity anomaly is well defined, the excess or missing mass can be computed directly from the gravity data, although the distribution of that mass in the subsurface cannot be defined. The non-uniqueness of the potential field can only be overcome by assuming that the structure fits some pre-conceived understanding concerning its density, shape, and position.

In finding a density model for a gravity anomaly the details of the assumptions will influence and often determine the solution. In effect, any solution that satisfies the restriction concerning the maximum depth to the top of the structure will be determined principally by the assumptions concerning the nature and shape of the density structure. The most common assumptions are based on well-understood shapes for geologic structures. Assumptions appropriate for a flat and shallow sedimentary basin differ greatly from those appropriate for the thin vertical plane of an intrusive dike, and the gravity anomalies would also have significantly different shapes. On the other hand, a sphere at depth might duplicate the approximate shape of an anomaly from a symmetrical shallow basin, but such a model would be an unrealistic solution where the surface geology clearly shows the existence of a basin. An obvious constraint on any model is that the modeled structure should not extend above the surface, a contradiction often created with the application of models that are too simple.

The density of the structure is, perhaps, the most limiting and important constraint. However, in modeling structures it is the density contrast and not the absolute density that controls the shape and magnitude of the anomaly. As will be shown, adding an arbitrary constant to all densities will only shift the reference line for modeled gravity data and the reference line or zero-anomaly line is a consequence of regional structures and the reduction equations. The density contrast is a strong constraint because the allowable density contrast of rocks in most geologic environments generally falls within a narrow range. The determination of an appropriate density model for a gravity anomaly is an exercise in using the gravity anomaly as a constraint. One generally starts with a model based on geological information. Although the gravity anomalies can constrain the depth and magnitude of the density model, the shape and density contrast is principally determined by our assumptions concerning the geologic cause of the anomaly. Realistic modeling of a structure begins with knowledge of prevalent rock densities and geologic structures.

The direct problem in potential theory, which is the topic of this chapter, is the computation of an anomaly from an assumed shape for the distribution of densities. The inverse problem, which is the estimation of the density structure from the gravity anomaly, incorporates the direct problem as an essential component. The simplest form of gravity data inversion is to start with an initial model and iteratively adjust the model to fit the data. The anomaly computed from a trial model is compared to an observed anomaly and the misfits in the comparison are used with geologic constraints to modify the original model. The modified model then becomes the new trial model and the comparison repeated with the objective of improving the fit of the modeled anomaly to the observed anomaly. Hopefully, the process is convergent. When an acceptable fit is obtained, one that only contains easily explainable differences, the density model becomes one of an infinite number of possible models that fit the anomaly. Again, multiple acceptable solutions are a consequence of the non-uniqueness of potential data. However, when realistic geologic constraints are used along with any available borehole data or other a priori information, the models can represent valid interpretations of the subsurface density structure.

6.2 Densities of rocks

The most relevant constraints on a density model are the densities of the rocks that make up the structure. These density values will allow estimation of the density contrasts. The traditional standard density for the Bouguer reduction in regional gravity surveys is 2.67 g/cm^3, a value consistent with the average density of granitic rocks. The standard reduction was originally based on the assumption that the granitic crust is principally a mixture of quartz at 2.66 g/cm^3 and feldspar at 2.67 g/cm^3 with minor contributions from denser minerals. Using a standard density in the gravity reduction of regional gravity data is more important than using a correct value because a standard density facilitates a direct comparison of Bouguer anomaly maps. However, the resulting maps may contain artifacts created by the difference between the assumed and actual density of the near-surface rocks. The most common manifestation of a deviation of the near-surface density from the reduction density is a correlation of anomaly details with topography. Near-surface sedimentary rocks, such as those common in coastal plain areas, might have a density closer to 2.0 g/cm^3, a density significantly lower than the reduction density. The crystalline rocks in an area of high-grade metamorphism can have densities on the order of 2.9 g/cm^3, a value that is significantly more dense than the rocks in a sedimentary basin. Although rock densities in these areas may differ from the standard Bouguer reduction density, the difference rarely exceeds more than 25 percent. Unconsolidated materials (soil, muck, and fill) often have a much lower density than the underlying rock and can have a density contrast of 100 percent or more. Choosing the correct Bouguer density to account for topographic variations of these low-density materials is especially important in local microgravity surveys.

The exact density of a rock unit is difficult to measure in place and, consequently, estimates of its density come primarily from laboratory measurements of rock samples or estimates from its mineral content. Data from wells with density logs can contribute significantly to the understanding of the unit's densities. Also, downhole gravity measurements with a borehole gravimeter are an option for determining the vertical density profile. The shape and magnitude of a gravity anomaly can often be used as a constraint on the appropriate density contrast to be used in an analysis. The applicability of density measurements of field samples is often questionable because the easily accessible rocks are near the surface, weathered, and represent selected weathering-resistant portions of the rock mass. The density of rocks that have no pore space or voids can be estimated from the fractional volume of the minerals that make up the rock. Because most fractures and pore spaces close below depths of about 4 km, the density of rocks at depths below about 4 km can be estimated from their mineral composition if known. Samples for density determination can thus be obtained from nearby surface exposures or well cuttings. At shallow depths the density that is important in gravity anomalies is the bulk density, the average density of a rock including all pore space and included fluids. Sedimentary rocks, like dolomite and limestone that are subject to the solution of large and numerous cavities near the surface, may show negative density anomalies at shallow depths and positive density anomalies when the rock unit is at depths below areas subject to rock loss by solution. The bulk density may be either dry or fluid saturated (wet). The wet density is the density of the rock with all pore spaces filled with fluids. The bulk density of rocks near the surface and above the water table is usually the dry bulk density. In most non-arid regions, the bulk density of a rock at a shallow depth and below the water table should be the wet bulk density. The density log, based on Compton scattering of gamma rays, can give estimates of bulk density. The dependence of bulk density on porosity has made the density log a valuable tool in oil exploration.

A geologic unit that was formed in a unique environment and that had the same geological history may often have a characteristic density or narrow range in densities. However, units with the same geologic description can vary in density over a significantly wider range. In effect, the densities of geologic units in a study area may assume characteristic values, but these values may be difficult to measure or to find appropriate values in the literature because rocks in that type of geologic unit may have a wide range of values. Some relations can be used to improve the estimate of the density of a rock. The density of structures with the same geologic description can vary systematically with age, depth of burial, and degree of metamorphism. Rocks that have survived for longer time periods are more likely to be compact and tightly cemented, hence denser than younger rocks of the same type. Rocks that are at greater depths, or have experienced the pressures of greater depths, are also more likely to be denser than rocks that have remained close to the surface. Similarly, rocks of a higher metamorphic grade are likely to be denser than low-grade metamorphic rocks. Age, pressure, and metamorphism all compress a rock and destroy its

Table 6.1 Average densities of common sedimentary rocks (abstracted from Dobrin, 1976).			
Rock Type	Range	Average (dry)	Average (wet)
Peat and muck	0.5–1.3	–	1.2
Alluvium	1.5–2.0	1.54	1.98
Clay	1.3–2.6	1.7	2.21
Sandstone	1.6–2.76	2.24	2.35
Shale	1.56–3.2	2.1	2.4
Limestone	1.74–2.9	2.11	2.55
Dolomite	2.0–2.9	2.3	2.7

Table 6.2 Ranges of densities for igneous rocks after Telford and others (1990).					
Rock Type	Range	Average	Rock Type	Range	Average
Obsidian	2.2–2.4	2.30	Diorite	2.7–3.0	2.85
Rhyolite	2.3–2.7	2.52	Diabase	2.5–3.2	2.91
Dacite	2.3–2.8	2.58	Basalt	2.7–3.3	2.99
Andesite	2.4–2.8	2.61	Gabbro	2.7–3.5	3.03
Granite	2.5–2.8	2.64	Peridotite	2.8–3.4	3.15
Anothosite	2.6–2.9	2.78	Pyroxenite	2.9–3.3	3.17

porosity. For metamorphic and igneous rocks, the mafic rocks are typically more dense than the acidic rocks, for example 2.79 g/cm^3 versus 2.61 g/cm^3, respectively. For these reasons, a tabulation of densities can be useful but should not be strictly relied upon for modeling structures.

Sedimentary rocks are on average less dense than igneous and metamorphic rocks. For this reason sedimentary basins are characterized by negative Bouguer gravity anomalies that can be used to model the depth and shape of the basin. The contrast in density is important in defining the magnitude of the anomaly, rather than the absolute density of the sediments. In general, the densities of sediments cover a wide range and are difficult to characterize for the different types of sedimentary rocks. The average density varies with composition. The different rock and soil types in order of increasing average density are: organic soils (e.g. peat and muck), alluvium, clay, sandstone, shale, limestone, and dolomite (Table 6.1). The variation in density gives considerable overlap in values. The wide range in densities is primarily related to the porosity. The susceptibility of porosity to compression and cementation explains why sedimentary rocks are strongly influenced by age, depth of burial and geologic history. The effects of depth of burial are more pronounced on shale than on sandstones and limestone because shale is more susceptible to dehydration during compaction, viscous deformation, and low-grade metamorphism than the chemically more homogeneous sandstones and limestone.

Igneous rocks (Table 6.2) are generally denser than sedimentary rocks; but, there is considerable overlap among the extrusive igneous rocks. The extrusive igneous rocks,

Table 6.3 Ranges of densities for metamorphic rocks.					
Rock Type	Range	Average	Rock Type	Range	Average
Quartzite	2.5–2.7	2.60	Slate	2.7–2.9	2.79
Schist	2.4–2.9	2.65	Gneiss	2.6–3.0	2.80
Phyllite	2.7–2.7	2.74	Amphibolites	2.9–3.1	2.96
Marble	2.6–2.9	2.75	Eclogite	3.2–3.55	3.37

Table 6.4 Densities of miscellaneous materials.					
Rock Type	Range	Average	Rock Type	Range	Average
Ice	0.88–0.92	0.9	Sea Water	1.01–1.05	1.03
Coal	1.1–1.8	1.32	Petroleum	0.6–0.9	–
Chalk	1.5–2.6	2.00	Rock Salt	2.1–2.6	2.22
Oxides	3.5–9.7	–	Ilmenite	4.3–5.0	4.67
Magnetite	4.9–5.2	5.12	Hematite	4.9–5.3	5.18
Sulfides	3.5–8.0	–	Pyrite	4.9–5.2	5.0

because they are formed in the low-pressure environment of the free surface, may be porous and may contain void spaces, significantly lowering their density when compared to the chemically equivalent intrusive igneous rocks. Mafic igneous rocks are denser than acidic igneous rocks. The range in densities of metamorphic rocks (Table 6.3) differs little from the range in densities of intrusive igneous rocks. For low-grade metamorphism, the metamorphic rocks are typically denser than their equivalent sedimentary rock. High-grade metamorphic rocks are often found in contact with intrusive igneous rocks with which they share a common geologic history. In regions of sedimentary and metamorphic rocks the density structure can be varied and complex at all scales. The density contrasts found at the scale of mineral grains can extend to the scale of meters. In sediments, areas of uniform density are limited to the more massive units associated with rapid deposition. Although the variations in density can be great across bedding planes, often the variations in average density are slight parallel to the bedding. In metamorphic terrains, the uniformity of density is determined by the character of the source rock. The influence of density variations at small scales is to add noise to the observed gravity, because the scale of the density variations is generally smaller than gravity data station separation. In contrast, igneous intrusive derived from magma may be relatively uniform in density and their density contrast with the surrounding host rock may be useful for modeling the geologic unit.

Most geologic units are complex mixtures of rock types for which average densities must be found from a combination of gravity anomaly values, measured densities of field samples, and estimates from the literature. However, some miscellaneous materials occur in large enough units to be significant in measurements of the gravity field (Table 6.4).

Some of these like sea water, the oxides and sulfides have densities that are well defined by their chemical composition. When these miscellaneous materials are found in massive units large enough to be detected by gravity measurements, they may be of significant economic interest.

6.3 Nettleton's method for determination of reduction density

Nettleton's method (Nettleton, 1939) of determining the reduction density from gravity data can, in some regions, give reasonably satisfactory estimates of an appropriate average near-surface density for use in Bouguer reduction when the influence of topography needs to be minimized. For the complete Bouguer reduction, a topographic correction should be included using an average density for near-surface rocks. Without the terrain correction in a valley, the Nettleton reduction density would be higher than the actual density. Hence, the actual rock density may differ from the best fit obtained by Nettleton's method when applied to data without the terrain correction.

The use of an appropriate reduction density will remove a correlation of observed gravity with topography. In Nettleton's method, a gravity profile over a study area, in which there are significant sharp variations in elevation, is reduced by using a range of Bouguer reduction densities. If the average density at the surface were uniform, the Bouguer anomaly using the density of the near-surface rocks should give anomalies with minimal correlation with the topography. The reduction density with a minimum correlation with elevation would give anomalies most representative of the underlying geologic structure. The profile in Figure 6.1 exhibits stream cuts at 3 km and 9 km. Both represent a 20-m deep cut in the surface rocks and both have a distinct signature in the Bouguer anomalies. Examination of Figure 6.1 indicates that the expressions of the stream cuts are minimized at a reduction density of approximately 1.75 g/cm^3. This is likely a good reduction density for the near-surface unconsolidated Coastal Plain sediments at this location. The actual density could be less because a topographic correction would be on the order of 0.1 mGal for these stream cuts and would lower the density required to minimize the influence of topography. Nettleton's reduction density is appropriate for the interpretation of anomalous density structures, but should be used with caution, or with further analysis including a topographic correction if estimating surface rock density.

Nettleton's method for determining the reduction density, or equivalently an estimate of the density of the near-surface rocks, can be generalized to larger areas to map the reduction density. In order to map reduction density, the elevation in smaller sub-areas is compared to the local Bouguer anomaly, which is the Bouguer anomaly with regional trends removed. To obtain the local Bouguer anomaly, a plane or smoothed version of the Bouguer anomaly is subtracted from the total Bouguer anomaly. The size of the sample area would be dependent on the topographic relief, with the objective of capturing significant short-length variations. The covariance of elevation and Bouguer anomaly is then computed as a function of reduction density. Figure 6.2 shows the covariance of the elevation and

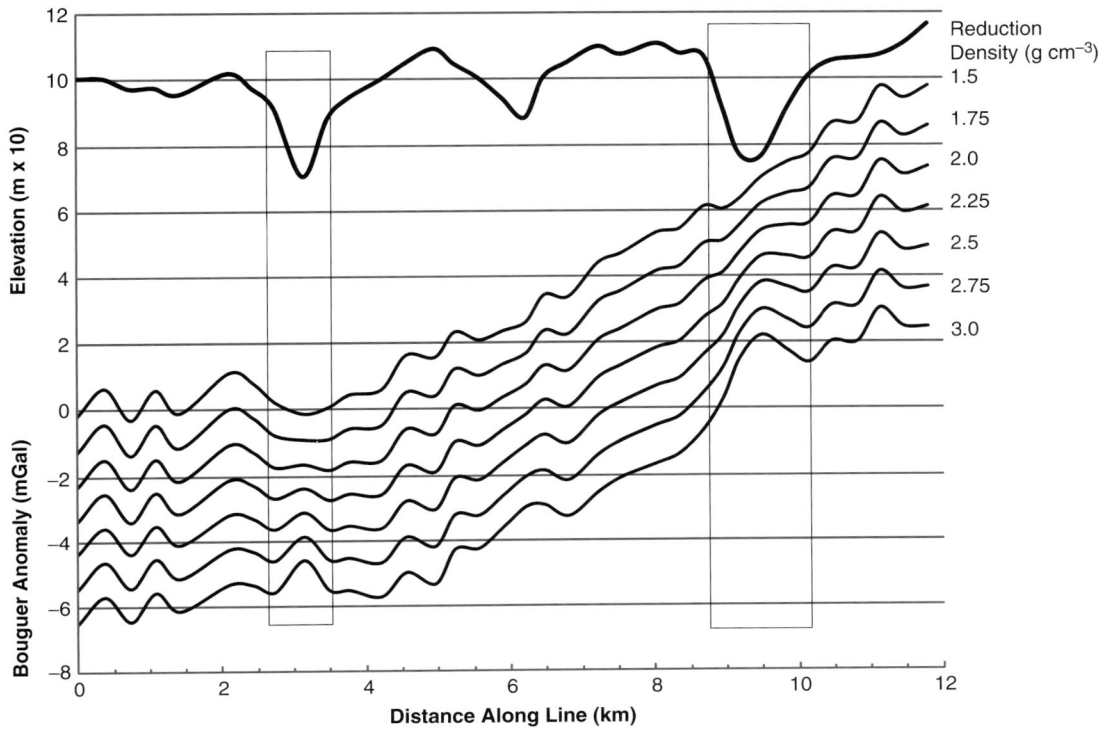

Figure 6.1 Application of Nettleton's method to a 12-km line of data in Georgia Coastal Plain sediments. The heavy line at the top is elevation, the lower lines are Bouguer anomalies computed by using reduction densities from 1.5 to 3.0 g/cm^3.

Bouguer anomalies computed as a function of reduction density for the topographic feature at 9 km. The covariance is minimized at a reduction density of 1.75 g/cm^3, thus confirming the reduction density found by inspection of Figure 6.1.

Nettleton's method cannot be used in many areas because the topography is influenced by changes in the rocks making up the geologic structure. When the geology and topography change at the same position along a line, the Bouguer anomaly may also change at that point. In these instances the reduction density that removes the correlation of topography with the Bouguer anomaly will be the one that minimizes the combined effects of the geologic structure and the topography. These misleading reduction densities can deviate as much as ±6.0 g/cm^3 from the density of the near-surface rocks. The best estimates of near-surface densities are obtained in areas where the topography is within a single geologic unit and where variations in topography are caused by sharp and recent stream erosion. The correlation that best smoothes the anomalies associated with short-wavelength topographic features usually gives the best values for average surface density. The long-wavelength topographic features are more likely to correlate with geologic structures and give unreliable estimates for a reduction density.

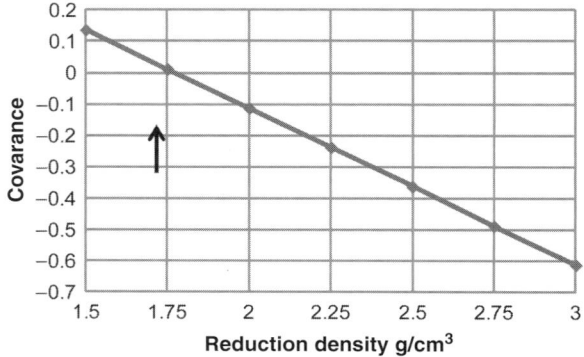

Figure 6.2   Covariance of Bouguer anomalies and elevation for different reduction densities.

6.4 Seismic velocity as a constraint for rock density

Seismic velocity is often helpful in constraining density. Although wave-equation solutions give velocities that are proportional to the square root of the appropriate elastic constants and inversely proportional to the square root of density, measurements of rock velocity and rock density suggest that seismic velocity typically increases with increases in density. On the basis of empirical data, Birch (1961) introduced *Birch's Law*, a linear relation between seismic velocity and density. This relation has been used in seismic tomography to relate velocity anomalies to density anomalies (Lees and VanDecar, 1991, Kaufmann and Long, 1996). *Birch's Law* is expressed in the form

$$v = a(M) + b\rho, \qquad (6.1)$$

where v is the compressional velocity, ρ is density and $a(M)$ and b are constants. Of significance is the relation between $a(M)$ and the mean atomic weight, M. The linear relation fits measurements for many crustal and mantle rocks. By ignoring the minor changes in average composition of the crust, and hence assuming the mean atomic weight is constant, Birch's relation may be differentiated to relate changes in velocity to changes in density as

$$\Delta v = b\Delta\rho, \qquad (6.2)$$

where b is the slope. The relation varies only slightly with depth. The more general Nafe–Drake relation (Nafe and Drake, 1963) which is based on laboratory measurements of rock velocity and density, exhibits a nonlinear relationship. A slope of around 6.15 corresponds to velocities between 3.5 km/s and 6.4 km/s, appropriate for the shallow crust. However, any use of a velocity versus density relation in interpretation needs to be calibrated with rocks that are typical for the study area and depth of the anomalous structure. An empirical density versus velocity relation assimilating earlier relations was proposed by Brocher (2005) and is summarized in Figure 6.3. When applied to sedimentary rocks at shallower depths, the slope of the curve differs for separate suites of rocks and will depend on their average

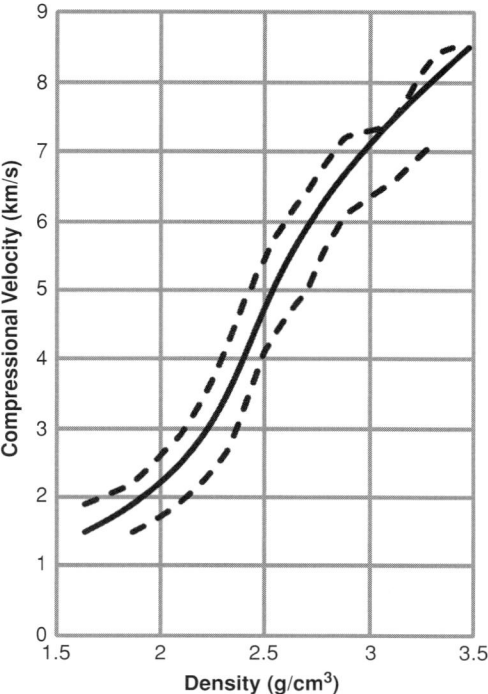

Figure 6.3 Relations between density and seismic velocity.

density and depth of burial. For sedimentary rocks Gardner et al. (1974) have demonstrated that the density follows a relation of the form

$$\rho = aV^m, \tag{6.3}$$

where V is velocity in m/s, ρ is the density and (a, m) are constants determined by the observed values of density and velocity. Gardner et al. (1974) gave default values of 0.31 and 0.25 for a and m respectively for brine-saturated sedimentary rocks. However, values obtained in their study area for shale were statistically different from those for sands.

6.5 Gravity anomalies from spherical density structures

The attraction of a sphere represents the simplest and most useful three-dimension approximation to a three-dimensional structure. At distance, the anomalies from all compact structures asymptotically approach the attraction of a sphere of the same mass with its center at the center of mass of the structure. In effect, the difference between the attraction of a sphere and a compact mass like a cube, tetrahedron, or amoebic blob, decreases rapidly from the center of mass of the structure. Differences are well within the error of most gravity measurements at a distance equivalent to a few diameters of the structure. Consequently, any complex or extended three-dimensional structure could be modeled by breaking the structure into distinct compact segments that may be approximated by spheres. The major constraints to be satisfied are that the density contrast of each sphere should be reasonably consistent with the geology and that the spheres should have a total volume that matches the volume of the structure causing the anomaly.

The attraction of a sphere in the Earth's field is from equation (1.36) is

$$a_z = \frac{\partial V}{\partial z} = \frac{4}{3}\pi a^3 G \Delta\rho \frac{(z - \zeta)}{\left((x - \xi)^2 + (y - \eta)^2 + (z - \zeta)^2\right)^{3/2}}, \tag{6.4}$$

where the anomalous mass of the sphere is given by

$$M = \frac{4}{3}\pi a^3 \Delta\rho \tag{6.5}$$

and the center of the sphere is at (ξ, η, ζ).

The mass of the sphere is proportional to the product of the density (or density contrast for anomalous mass) with the cube of its radius. The coupling of density with a dimension is an example of the non-uniqueness of potential data. Any combination of radius and density that gives the same total mass M will give the same curve for the anomaly. Hence, the constraint provided by reasonable densities and dimensions for the sphere determines the model once the shape of the anomaly is matched. The coupling of density with a dimensional component is inherent in all models used in computing anomalies. The observation that the attraction is equivalent to the total mass concentrated at the center of mass of the sphere is common to many symmetrical shapes, an observation that makes the sphere a good approximation for an arbitrarily shaped object observed at some distance.

The anomaly for a sphere is symmetrical about the center of the sphere. Hence, in Figure 6.4 only half of the anomaly for a sphere is shown. If we write Eq. (6.4) in cylindrical coordinates (r, θ, z) with the z-axis by convention taken positive down and if we place the origin on the surface above the center of the sphere, then on the surface, $z = 0$, Eq. (6.4) becomes

$$a_z(r) = \frac{4}{3}G\pi a^3 \Delta\rho \frac{\zeta}{\left(r^2 + \zeta^2\right)^{3/2}}, \tag{6.6}$$

where the symmetry of the anomaly eliminates the dependence on direction.

The half-width of the anomaly for a sphere can be used to estimate the maximum depth to the center of a structure's anomalous density. For this to be useful, the structure only needs to approximate a sphere. The half-width is the distance from the maximum value, taken over the center of the sphere, to the distance where the anomaly is half the difference between the maximum value and the gravity value corresponding to a regional trend line. An expression for the half-width can be found by equating half the maximum anomaly, evaluated at $r = 0$, to the expression for the anomaly from Eq. (6.6) evaluated at the radius for the half-height

$$\frac{1}{2}a_z(r = 0) = \frac{1}{2}\frac{\frac{4}{3}G\pi a^3 \Delta\rho}{\zeta^2} = \frac{\frac{4}{3}G\pi a^3 \Delta\rho\zeta}{\left(r_{1/2}^2 + \zeta^2\right)^{3/2}}, \tag{6.7}$$

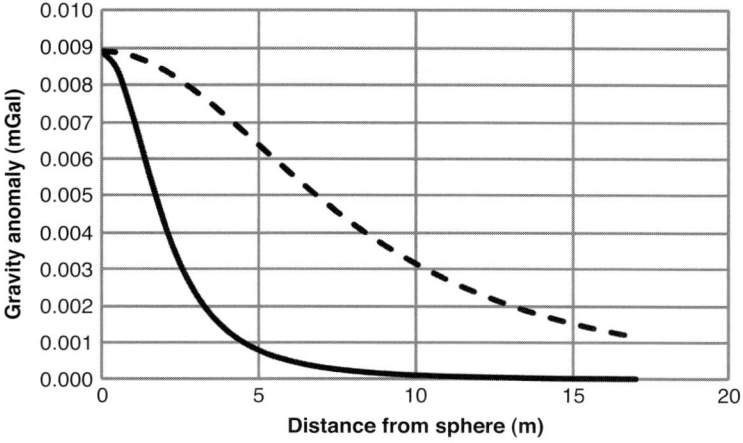

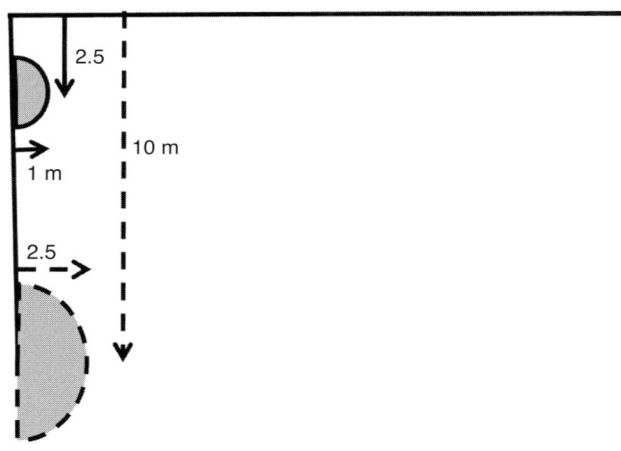

Figure 6.4 Anomaly from a sphere, showing the half-width and effect of depth.

which allows a solution for the depth ζ as a function of the half-width

$$\left(r_{1/2}^2 + \zeta^2\right)^{3/2} = 2\zeta^3 \tag{6.8}$$

$$1.305 r_{1/2} = \zeta. \tag{6.9}$$

The half-width relation can be used for any three-dimensional structure that is compact and nearly spherical in shape. The shape of the anomaly of a sphere places important constraints on the interpretation of shallow structures. If the station spacing is comparable to the half-width of an anomaly, then the station spacing will not provide enough information to define the shape of the anomaly. In order to obtain a meaningful model for a sphere, the center of the sphere should be at a depth greater than the station spacing. Depths shallower than the station spacing in essence become noise in the data.

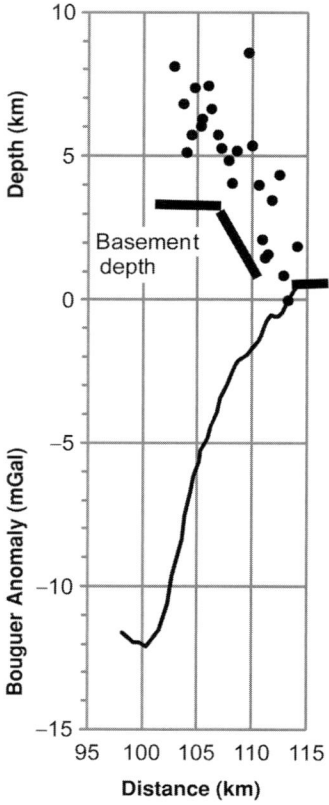

Figure 6.5 Automatic depth estimates from equations for a sphere.

The anomaly for a sphere can be used as a guide in finding a general relation for the maximum depth to the center of anomalous density structures. For example, if we compute the ratio of the magnitude of the gravity value to the radial gradient for the anomaly from a sphere, the maximum value occurs at a distance equal to the depth to the center. For three-dimensional anomalies in general we can write the maximum depth, h, to the center of the anomaly as

$$h \leq \frac{3}{2} \max \left| \frac{\Delta g}{\frac{\partial \Delta g}{\partial r}} \right|. \tag{6.10}$$

Because a sphere is the most concentrated form of density anomaly, this relation holds for any shape of anomaly approximating a sphere.

As an example, Eq. (6.10) is applied to the maximum depth estimates along a line crossing a Triassic basin (Figure 6.5). The depth of the basin is known to be 2.7 km. At the southern edge the sediments, which overlay the basin and metamorphic crustal rocks, are approximately 0.4 km deep. The anomalies at the southern end relate to a positive density anomaly in the metamorphic crustal rocks. The negative anomaly to the north is caused by the Triassic basin. To obtain this interpretation, Eq. (6.10) was applied to a 5-point moving average of the data. The moving average removes noise that is related to near-surface topography and meter reading uncertainty. The smoothed values were projected onto a straight line perpendicular to the structure. For interpretation of depth, the values closest to points where the gradient is the greatest should give the most reliable estimates of maximum depth. Values computed for positions where the gradient is slight and not at its maximum give anomalously large and unreliable estimates of maximum depth and are off the graph.

Relations comprising the slope and magnitude of local anomalies can be useful if interpreted with consideration of the noise in the data. Generally, the precision is not as robust as modeling the anomaly with reasonable constraints. Fitting a model to the data and using the magnitude of the modeled anomaly and slope would give the best solutions for depth. Noise will give false shallow depths so care must be taken to utilize well-defined anomalies. Also, if computing the derivative along a line, changes in the direction of the line can introduce rapid changes in the derivative and give false estimates of maximum depth. In the case of lines that are not straight, the data should be projected onto a straight line perpendicular to the contours. But even with lines projected to a straight line, gradients perpendicular to the line can give false shallow depths.

Downward continuation is an effective way to determine the depth of a causative density anomaly. By using finite-difference downward continuation or the Fourier transform method, the downward continued gravity anomalies become unstable at depths corresponding to the center of mass of distinct anomalies (see Figure 5.13). The estimates of depth so obtained will be the maximum possible depth to the top of the structure. The Fourier transform technique for downward continuation provides an alternate way to evaluate depth to the source of the anomaly. Because the attraction of a sphere can be expressed as the attraction of a point mass containing all the mass of the sphere, the density structure for a sphere is equivalent to an impulse at the center of the sphere. Hence, the gravity anomaly continued down to the center of the sphere is a simple impulse. Then Eq. (5.59) can be written as

$$\mathcal{F}\{\Delta g(x, y, z)\} = e^{-\left(p^2+q^2\right)^{1/2}z}\mathcal{F}\{\Delta g(x, y, 0)\}. \tag{6.11}$$

By taking the log of Eq. (6.11) and recognizing that the Fourier transform of the impulse is a constant, Eq. (6.11) can be written as

$$\log\left(\mathcal{F}(\Delta g(x, y, z))\right) = C - \left(p^2 + q^2\right)^{1/2}z = C - |k|\,z. \tag{6.12}$$

Equation (6.12) shows that the logarithm of the Fourier transform of the gravity anomaly is proportional to the product of the wavenumber and depth. By taking the derivative of Eq. (6.12) with respect to $|k|$, a relation for the depth can be written as

$$-z \sim \frac{\Delta\log\left(\mathcal{F}\{\Delta g\,(x, y, 0)\}\right)}{\Delta k}. \tag{6.13}$$

As an example, Figure 6.6 shows the log of the Fourier transform for a short line of gravity data in the South Carolina Coastal Plane where the sediments are a little over 1 km

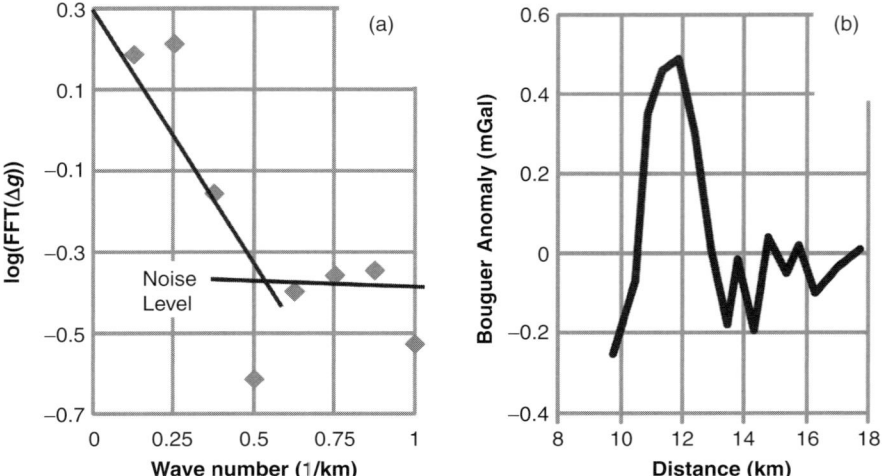

Figure 6.6 Fourier transform method for determination of depth to an anomaly. (a) Log Fourier transform of gravity data versus wavenumber. (b) Trace used in this computation.

thick. The slope of the line agrees with a 1.2-km depth to the basement. Also, Figure 6.6 shows a typical noise level for the larger wavenumber (shorter wavelengths) caused by near-surface uncorrelated anomalies and noise in the data. A flat noise line is consistent with very shallow sources for these wavenumbers.

6.6 The attraction of a thin rod

The attraction of a thin rod (or dipping line source) is an example of a model for structures that are not symmetric about a point. For a thin rod, the mass may be assumed to be concentrated along the axis of the rod. This approximation is valid so long as the radius of the rod is a factor of 4 less than the distance from the ends of the rod to the observation point. The derivation starts by defining the attraction of an infinitesimal length of the rod. The infinitesimal element shown in Figure 6.7 is given a cross-sectional area of A and a length of dl. The anomalous mass of the infinitesimal element is then

$$dm = \Delta \rho A dl. \tag{6.14}$$

The attraction of this infinitesimal element at distance will approach that of a point mass, and its vertical component at the surface ($z = 0$) is given as in Eq. (6.4) as

$$\delta g = G \Delta \rho A dl \left[\frac{z - \zeta}{\left((x - \xi)^2 + (y - \eta)^2 + (z - \zeta)^2\right)^{3/2}} \right] = G \Delta \rho A \frac{\zeta}{d^3} dl. \tag{6.15}$$

In order to derive a general equation, we tilt the axis of the rod an angle α from the horizontal. Also, without loss of generality, we orient the strike of the rod parallel to the

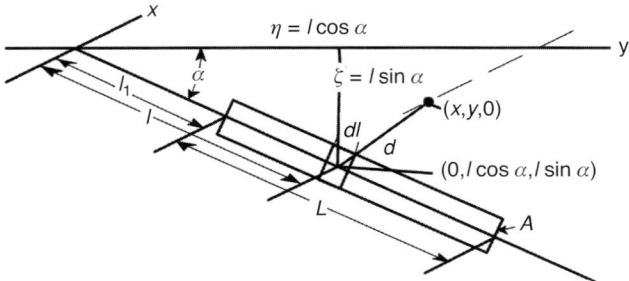

Figure 6.7 Reference geometry for the attraction of a dipping line source.

y-axis as shown in Figure 6.7. After computing the gravitational attraction, the values can be rotated to provide the anomaly in any orientation.

The distance d from the infinitesimal element to the observation point can be obtained from the locations of the infinitesimal element $(0, l \cos \alpha, l \sin \alpha)$ and observation point $(x, y, 0)$ and is given by

$$d^2 = \left(x^2 + (y - l \cos \alpha)^2 + (l \sin \alpha)^2\right)^{1/2}. \tag{6.16}$$

The value of gravitational attraction is computed for a rod starting at $l = l_1$ and ending at $l = l_1 + L$. The total attraction of the thin rod is the integral of the infinitesimal elements over the length of the rod

$$\delta g = G \Delta \rho A \sin \alpha \int_{l_1}^{l_1+L} \frac{l\,dl}{\left(x^2 + y^2 - 2yl \cos \alpha + l^2\right)^{3/2}}. \tag{6.17}$$

The integral in Eq. (6.17) can be put into a form found in standard integral tables to give

$$\delta g = 2G \Delta \rho A \sin \alpha \left[\frac{-2y \cos \alpha + 2\left(x^2 + y^2\right)}{\left(4y^2 \cos^2 \alpha - 4(x^2 + y^2)\right)\left(x^2 + y^2 - 2yl \cos \alpha + l^2\right)^{1/2}} \right]_{l_1}^{l_1+L}. \tag{6.18}$$

Equation (6.18) is useful for modeling cylindrical structures that are dipping at significant angles. For a slight dip, where α is near zero or near π, the expression approaches the two-dimensional equations that are derived below.

As an example, the negative anomaly for a tunnel, approximated as a rod with a 4-m cross-sectional area and negative density contrast of -2.0 g/cm^3 is shown in Figure 6.8. The tunnel starts at 4 m from the origin and dips down at 30 degrees. Notice the maximum anomaly is above the start of the tunnel, at its closest approach to the surface and decreases and widens where the tunnel gets deeper and in the direction of the dip of the tunnel. The maximum anomaly is on the order of -0.02 mGal but would be larger for larger cavities. Microgravity surveys (Butler, 1984) are useful in detecting similarly sized tunnels and cavities.

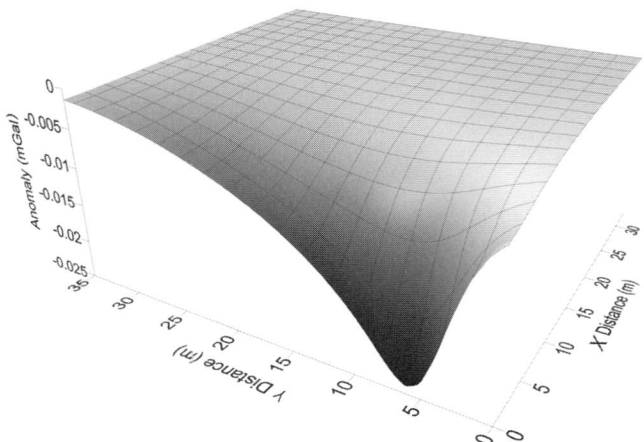

Figure 6.8 Anomaly from a 4 square meter cross section dipping 30 degrees. Tunnel starts at 5 m and extends to 29 m. See Figure 6.7. The maximum anomaly is at the shallow end with an elongation toward the deeper end.

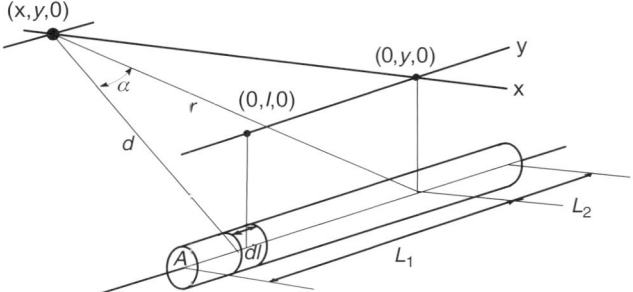

Figure 6.9 Reference geometry for a horizontal line source.

6.7 Attraction of a horizontal cylinder of finite length

The attraction of a horizontal cylinder of finite length may be computed as a simpler case of the attraction of a dipping thin rod. The horizontal cylinder is often the best approximation for karst conduits and dissolution zones, often the targets of microgravity surveys. For this derivation the rod is oriented parallel to the y-axis at a depth of ζ. Again, there is no loss in generality by orienting it parallel to the y-axis because the resulting anomaly can be rotated to any orientation. Given an incremental length of the rod, dl, located at $(0, y - l, \zeta)$, the mass of this incremental element is the product of its area, A, and the incremental length, dl. The vertical component of attraction is given from Eq. (6.15) and Figure 6.9 by

$$\delta g = GA\Delta\rho \frac{\zeta dl}{d^3}.$$

$$(6.19)$$

The distance d may be expressed in terms of its projection onto the perpendicular distance from the rod, r, as $d = r/\cos\phi$. Also, the incremental length dl can be expressed in terms of the incremental angle $d\phi$ from the relation

$$dl = \frac{d}{\cos\phi}d\phi = \frac{r\,d\phi}{\cos^2\phi}, \tag{6.20}$$

giving the attraction of an element

$$\Delta g = GA\Delta\rho\zeta\left(\frac{r\,d\phi}{\cos^2\phi}\right)\left(\frac{\cos^3\phi}{r^3}\right) = GA\Delta\rho\zeta\cos\phi\frac{d\phi}{r^2}. \tag{6.21}$$

This expression may be evaluated for a finite length of rod that extends from $y - L_1$ to $y - L_2$ parallel to the y-axis by integrating over the angle ϕ

$$\Delta g = \frac{GA\Delta\rho\zeta}{r^2}\int_{\tan^{-1}\frac{y-L_2}{r}}^{\tan^{-1}\frac{y-L_1}{r}}\cos\phi\,d\phi$$

$$= \frac{GA\Delta\rho\zeta}{r^2}\left(\frac{y - L_1}{\left((y - L_1)^2 + r^2\right)^{1/2}} - \frac{y - L_2}{\left((y - L_2)^2 + r^2\right)^{1/2}}\right). \tag{6.22}$$

In the case where the rod is centered about the value of $y = 0$, so that $L_1 = -L$ and $L_2 = L$. The equation simplifies to

$$\Delta g = \frac{2GA\Delta\rho\zeta}{\left(x^2 + \zeta^2\right)}\frac{L}{\left(L^2 + \zeta^2 + x^2\right)^{1/2}} \tag{6.23}$$

for measurements at the surface $z = 0$ for a rod with an axis at $x = 0$ and $z = \zeta$. For observations near the center of the rod, if L is greater than twice the depth, then the attraction is more than 90 percent of the attraction from an infinite rod and it would be more efficient to use the equations for a two-dimensional structure derived in the next section. On the other hand, in the limit as L approaches zero, and in conditions where the length of the rod is less than half the depth, Eq. (6.20) gives the attraction of a point mass of mass $A\Delta\rho L$. Hence, using the equation for a sphere in this case would be sufficient. Programs that utilize equations developed for rod lengths approximating the rod depth are referred to as 2.5-dimensional systems of modeling density anomalies.

If the rod length, L, in Eq. (6.23) is extended to infinity, the equation simplifies to the equation for a horizontal infinite cylinder

$$\Delta g\,(x, z = 0) = \frac{2GA\Delta\rho\zeta}{\left(x^2 + \zeta^2\right)}, \tag{6.24}$$

where the depth to the center is ζ and the cross-sectional area is A. The shape of the anomaly for a horizontal rod is similar to that for a buried sphere, but the half-width is different. If half the magnitude at $x = 0$ is set equal to the width at the half-amplitude, a relation between the depth ζ and half-width can be found,

$$\frac{\Delta g \, (x = 0, z = 0)}{2} = \frac{2GA\Delta\rho}{2\zeta} = \Delta g \, (x_{1/2}, z = 0) = \frac{2GA\Delta\rho\zeta}{\left(x_{1/2}^2 + \zeta^2\right)}$$

$$2\zeta^2 = \left(x_{1/2}^2 + \zeta^2\right) \tag{6.25}$$

$$x_{1/2} = \zeta.$$

For the horizontal cylinder the depth to the center is equal to the half-width. Also, the anomaly for a horizontal cylinder is inversely proportional to its depth, whereas for a sphere the amplitude is inversely proportional to the square of its depth.

6.8 The two-dimensional potential

Many density structures are elongated in one dimension. When their length is more than twice the distance to the observation point, they may be considered two-dimensional. Such structures allow the use of the two-dimensional or logarithmic potential in computing the gravitational attraction. Two-dimensional formulae for attraction are used extensively in modeling linear structures such as faults, mountain ridges and continental margins. However, the derivation of the two-dimensional potential requires some special consideration. For three-dimensional density anomalies with finite mass, we set the potential at infinity to be zero by convention. Two-dimensional bodies, in theory extend to infinity and thus do not have finite mass. Consequently, the potential is not bounded. The derivation for the gravitational attraction starts with the fundamental equation for the potential

$$V (x, y, z) = G \iiint \frac{\rho \, (\xi, \eta, \zeta) \, d\xi d\eta d\zeta}{\left((x - \xi)^2 + (y - \eta)^2 + (z - \zeta)^2\right)^{1/2}}. \tag{6.26}$$

The derivative of Eq. (6.26) with respect to z gives the vertical component of the attraction. To develop the two-dimensional attraction, we note that the density does not vary in the y direction and may be written as a function of only ξ and ζ. The integral is taken along the y-axis for mass distributed from $\eta = -\infty$ to $\eta = +\infty$,

$$V (x, z) = G \iint \rho \, (\xi, \zeta) \, d\xi d\zeta \int_{\eta=-\infty}^{\eta=\infty} \frac{d\eta}{\left((x - \xi)^2 + (y - \eta)^2 + (z - \zeta)^2\right)^{1/2}}, \tag{6.27}$$

leaving the integration over the other two dimensions as an integration over the area in the x and z plane. In order to evaluate the integral over η, the integral is first evaluated for finite distance L and then the limit is taken as L approaches infinity. It is convenient to use an infinite rod that is symmetric about the origin, as shown in Figure 6.9, and compute the integral from the origin with $y = 0$ to L and then compute the limit as L goes to infinity. Also for simplicity, we will use, r, the distance to the rod and let

$$r = \sqrt{(x - \xi)^2 + (z - \zeta)^2}, \tag{6.28}$$

and write the integral as

$$V(x, z) = 2G \iint \rho(\xi, \zeta) d\xi d\zeta \lim_{L \to \infty} \int_0^L \frac{d\eta}{(r^2 + \eta^2)^{1/2}}. \tag{6.29}$$

From integral tables the integral over η can be evaluated as

$$V(x, z) = 2G \iint \rho(\xi, \zeta) d\xi d\zeta \lim_{L \to \infty} \ln\left(\eta + (r^2 + \eta^2)^{1/2}\right)\Big|_{\eta = 0}^{\eta = L} \tag{6.30}$$

and thus

$$V(x, z) = 2G \iint \rho(\xi, \zeta) d\xi d\zeta \lim_{L \to \infty} \ln\left(\frac{L + (r^2 + L^2)^{1/2}}{r}\right). \tag{6.31}$$

In the limit as L goes to infinity, the potential goes to infinity, which is not realistic. The mass of a two-dimensional infinite body is not finite and to circumvent this, the definition of the potential must be modified for two-dimensional structures. The modification chosen is to add a constant that makes the potential go to zero at unit radius. Adding a constant will have no effect when taking the derivative to obtain the gravitational attraction. Hence, we subtract the value at $r = 1$ to get

$$V(x, z) = 2G \iint \rho(\xi, \zeta) d\xi d\zeta \lim_{L \to \infty} \left(\ln\left(\frac{(L^2 + r^2)^{1/2} + L}{r}\right)\right.$$
$$\left. - \ln\left(\frac{(L^2 + 1^2)^{1/2} + L}{1}\right)\right). \tag{6.32}$$

The limit as L goes to infinity in Eq. (6.32) yields the logarithmic potential

$$V(x, z) = 2G \iint \rho(\xi, \zeta) \ln\frac{1}{r} d\xi d\zeta. \tag{6.33}$$

The vertical component of attraction is obtained by differentiating with respect to z

$$\Delta g(x, z) = 2G \iint \frac{\rho(\xi, \zeta)(z - \zeta)}{(x - \xi)^2 + (z - \zeta)^2} d\xi d\zeta. \tag{6.34}$$

Although two-dimensional bodies do not exist, small contributions from distant portions of the anomalous mass allow their use when the length L is more than four times the depth, which corresponds to the condition that 90 percent of the anomaly is caused by the density anomalies within a distance of twice the depth. The next few sections will examine applications of Eq. (6.34) to two-dimensional structures of various shapes.

6.9 Vertical sheet

The attraction of simple two-dimensional structures can be obtained by the integration of Eq. (6.34) over the cross-sectional area of the structure. We can start by examining the

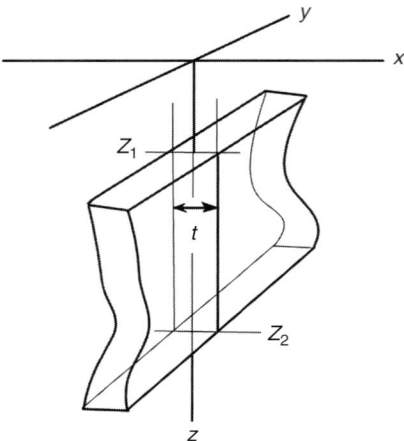

Reference geometry for a vertical sheet.

attraction of a thin sheet that can be computed by using the approximation that the sheet is infinitesimally thin. For a vertical sheet, such as a dike shown in Figure 6.10 where the thickness is relatively insignificant, Eq. (6.34) is integrated in the horizontal direction first using the mean value theorem, to give

$$\Delta g\,(x, z) = 2G \int_{z_1}^{z_2} \int_{\xi_0 - t/2}^{\xi_0 + t/2} \frac{\Delta \rho\,(\xi, \zeta)(z - \zeta)}{(x - \xi)^2 + (z - \zeta)^2}\,d\xi\,d\zeta. \tag{6.35}$$

This corresponds to the attraction of a vertical dike of thickness t with its top at depth z_1 and its base at z_2. For simplicity, the density is set to a constant and may be taken outside the integral. Furthermore, the thickness t must be set small relative to the depth z_1. Hence, when integrating ξ

$$\int_{\xi_0 - t/2}^{\xi_0 + t/2} f(\xi)d\xi \approx f\,(\xi_0) \int_{\xi_0 - t/2}^{\xi_0 + t/2} d\xi = tf(\xi_0) \tag{6.36}$$

over the dike thickness the expression for attraction varies slowly and the mean value theorem may be used to approximate the integral. In Eq. (6.36) the mean value, ξ_0, is taken at the center or axis of the dike. The attraction of a thin vertical sheet is

$$\Delta g\,(x, z) = 2G \Delta \rho t \int_{\zeta = z_1}^{\zeta = z_2} \frac{(z - \zeta)}{(x - \xi_0)^2 + (z - \zeta)^2}\,d\zeta, \tag{6.37}$$

which may be integrated to the form

$$\Delta g\,(x, z) = G \Delta \rho t \ln((x - \xi_0) + (z - \zeta)) \Big|_{\zeta = z_1}^{\zeta = z_2}. \tag{6.38}$$

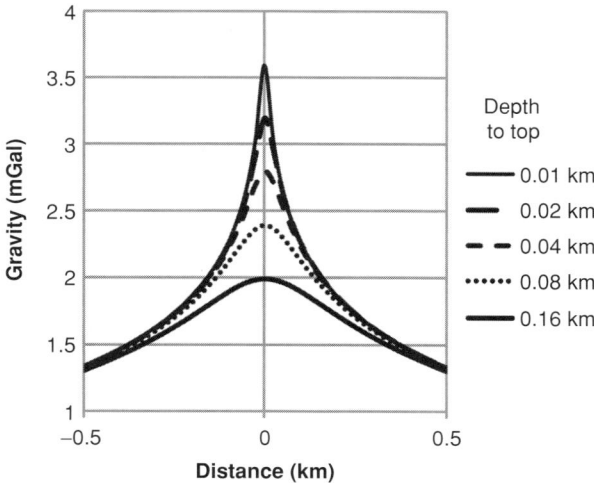

Figure 6.11 Dike anomaly with variable depth to top.

When evaluated at the surface, $z = 0$, for a vertical sheet located at $\xi_0 = 0$ Eq. (6.38) gives the simpler expression

$$\Delta g = 2G \Delta \rho t \ln \left(\frac{x^2 + z_1^2}{x^2 + z_2^2} \right). \tag{6.39}$$

The anomaly for a vertical sheet of thickness 5 m and density contrast of 0.1 g/cm^3 is shown in Figure 6.11. The top of the sheet varied from 160 m to 10 m and the base is extended to 5 km. The net effect of extending the sheet to greater depths is an increase in the regional positive anomaly, recognizing that a two-dimensional feature, when extended to infinity, has infinite mass. The principal identifying character of the anomaly is caused by the shallow portion of the sheet and is primarily determined by the depth to the top. Increasing the depth to the top smoothes and lowers the anomaly, whereas decreasing the depth sharpens the anomaly, and the mean approximation will be accurate so long as the requirement that $z_1 > t$ is maintained.

6.10 Horizontal half-sheet

The horizontal half-sheet shown in Figure 6.12 has considerable utility in providing equations for the anomaly of simple near-vertical faults and other structures that are easily approximated by the edge of a horizontal plate. For example, topography on the contact between basement rock and overlying sediments can be approximated by a half-sheet because only the uplifted portion is anomalous. As with the vertical sheet, when the thickness, t, is much smaller than the depth, ζ_0, we can use the mean value theorem to approximate the integration over the vertical dimension. The limits of the integration in

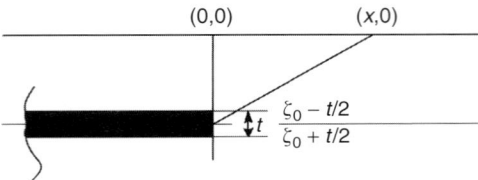

Figure 6.12   Reference figure for a horizontal half-sheet.

the vertical direction are half the thickness centered on the middle of the horizontal sheet. Hence, in this approximation the attraction of the horizontal plate of thickness t is approximated by an infinitesimally thin sheet located in the middle of the plate. The equation is

$$\Delta g\,(x,z) = 2G\Delta\rho \int_{\zeta_0-t/2}^{\zeta_0+t/2} \int_{\xi=-\infty}^{\xi=0} \frac{(z-\zeta)}{(x-\xi)^2 + (z-\zeta)^2} d\xi\, d\zeta. \tag{6.40}$$

For an anomaly measured at elevation z, and the mean value of the expression taken at ζ_0, the depth to the center of the flat plate, the variable of integration can be changed to $x - \xi$ to give

$$\Delta g\,(x,z) = -2G\Delta\rho t \int_{+\infty}^{x} \frac{(z-\zeta_0)}{(x-\xi)^2 + (z-\zeta_0)^2} d\,(x-\xi). \tag{6.41}$$

Equation (6.41) can be integrated directly and evaluated at the limits to give the equation

$$\Delta g\,(x,z) = -2G\Delta\rho t \left(\frac{\pi}{2} - \tan^{-1}\frac{x}{z-\zeta_0}\right). \tag{6.42}$$

In the limit as x approaches negative infinity $\Delta g = 2\pi G\Delta\rho t$, the equation for the Bouguer plate.

The depth of the simple fault can be estimated by using Eq. (6.42). At the surface $z = 0$ and the point where $x = \zeta_0$, the arc tan gives $\pm\pi/4$. These values correspond to values of the anomaly that are $\frac{1}{4}$ and $\frac{3}{4}$ of the maximum value. The distances from the middle of the anomaly to the points corresponding to these values indicate the depth. Simple fault anomalies at 5 different depths are shown in Figure 6.13. In Figure 6.13, the $\frac{3}{4}$ maximum values are 1.05 and the intersection of the curves along this line corresponds to their respective depths. Note also in Figure 6.13 the graph for a depth of 30 km is very broad and would be indistinguishable from a regional trend.

As an example of interpretation, Figure 6.14 shows the gravity profile taken in the northeast direction along the axis of a positive gravity anomaly near Bowman, South Carolina. A regional trend was removed before plotting. The curve corresponds to a depth of 0.6 km with the edge of the flat sheet at the position $x = -0.2$ km. With a suggested density contrast of 0.67 g/cm^3 the thickness would be approximately 110 meters.

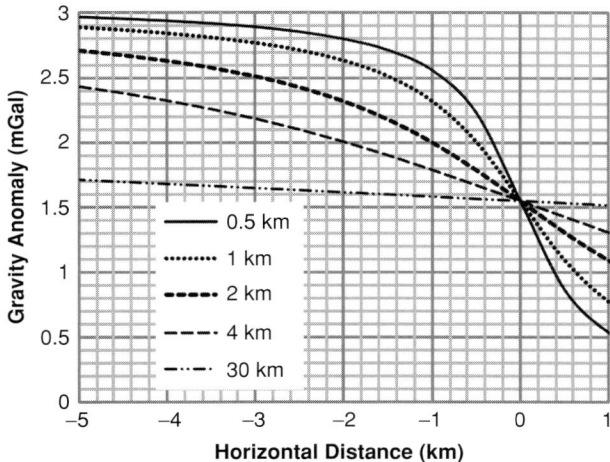

Figure 6.13   Anomaly for a simple fault with variable depth.

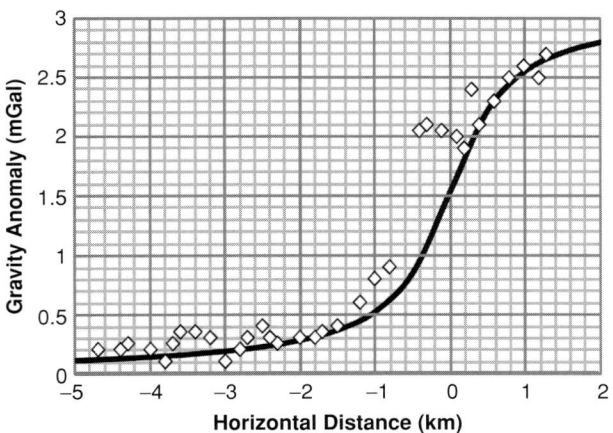

Figure 6.14   Anomaly for a simple fault (horizontal sheet) at a depth of 0.6 km.

6.11 Two-dimensional polygonal-shaped bodies

The two-dimensional expression for gravity anomalies can be generalized to a two-dimensional structure with an arbitrarily defined polygonal shape. Given a polygonal cross section as shown in Figure 6.15 we can consider the equation for only one side at a time. The expression is found for the attraction of a flat plate extending to infinity from the side of the polygon. One then calculates the attraction for each side sequentially progressing around the polygon in a clockwise direction. Contributions from the up-going sides outside the polygon will cancel out those from the down-going sides from the edge of the polygon to infinity, leaving only the attraction of the two-dimensional polygon.

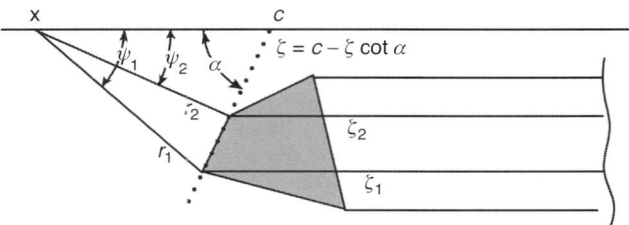

Figure 6.15 Reference figure for a two-dimensional polygon.

In order to set up the integral, the line defining one side is defined by the relation $\xi = c - \zeta \cot \alpha$ as shown in Figure 6.15. The integral is taken from the line defining the polygon edge to infinity and between the depths at the top and bottom of the sheet

$$\Delta g(x_1) = 2G\Delta\rho \int_{\zeta_1}^{\zeta_2} \int_{\xi=c-\zeta \cot \alpha}^{\xi=\infty} \frac{\zeta - z}{(\xi - x)^2 + (\zeta - z)^2} d\xi d\zeta. \tag{6.43}$$

Evaluating the integral for an observation on the surface, $z = 0$, and from the line representing the side of the polygon to $\xi = \infty$ gives the relation

$$\Delta g(x_1) = 2G\Delta\rho \int_{\zeta_1}^{\zeta_2} \left(\frac{\pi}{2} - \tan^{-1}\left(\frac{c - x - \zeta \cos \alpha}{\zeta}\right) \right) d\zeta. \tag{6.44}$$

The integral from ζ_1 to ζ_2 takes the form

$$\Delta g(x_1) = 2G\Delta\rho \left(\begin{array}{c} \frac{\pi}{2}\zeta + \zeta + \zeta \tan^{-1}\left(\frac{\zeta \cot \alpha - (c - x)}{\zeta}\right) \\ -\frac{1}{2}(c - x)\sin^2\alpha \ln\left((c - x - \zeta \cot \alpha)^2 + \zeta^2\right) \\ -(c - x)\sin\alpha\cos\alpha\tan^{-1}\left(\frac{\zeta - (c - x)\sin\alpha\cos\alpha}{(c - x)\sin^2\alpha}\right) \end{array} \right) \Bigg|_{\zeta_1}^{\zeta_2}. \tag{6.45}$$

Expressed in terms of the distance and angle from the origin to the corners of the polygon, Eq. (6.45) takes the form

$$\Delta g(x_1) = 2G\Delta\rho \left(\zeta_2\psi_2 - \zeta_1\psi_1 - (c - x)\left(\sin^2\alpha \ln\left(\frac{r_2}{r_1}\right) + (\psi_2 - \psi_1)\sin\alpha\cos\alpha\right) \right), \tag{6.46}$$

where the angles are also defined in Figure 6.15. In two-dimensional computation programs some form of Eq. (6.45) or (6.46) is used to define the anomalous density structure. Any two-dimensional structure can be modeled to any level of precision by increasing the number of polygons and the number of their sides. However, details that are smaller than four times the depth to the modeled feature are rarely warranted.

In generating a model to fit observed data, there is always a decision to be made concerning what portion of the anomaly is a regional trend and what portion is the target structure. The non-uniqueness of potential data does not help with this decision. Additional

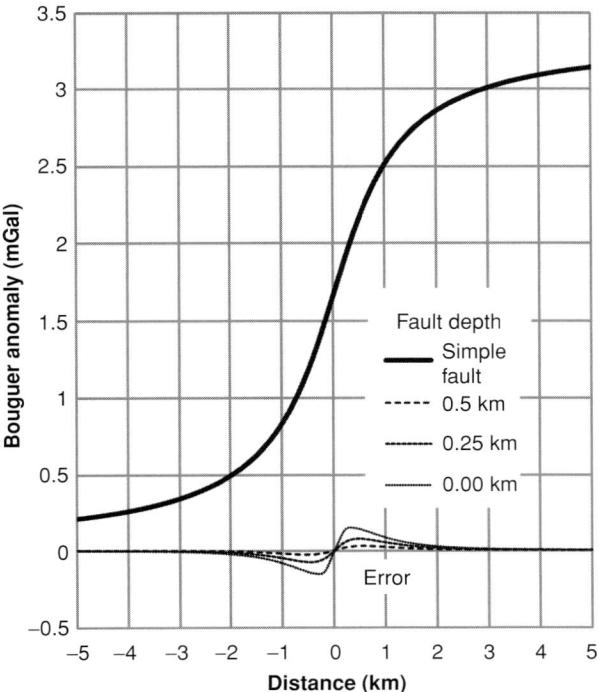

Figure 6.16 Error associated with the equation for a simple fault relative to the non-approximate fault model in two dimensions.

information is needed to force the modeled structure to be consistent with the geologic structures. For shallow structures, the regional trends are usually obvious and can be easily removed.

If the side of the polygon is vertical, where $\alpha = \pi/2$, Eq. (6.46) simplifies to

$$\Delta g\,(x_1) = 2G\,\Delta\rho \left(\zeta_2\psi_2 - \zeta_1\psi_1 - (c - x)\ln\left(\frac{r_2}{r_1}\right)\right). \tag{6.47}$$

Equation (6.42), the equation for a simple fault, is an approximation to Eq. (6.47). The amount of error depends of the thickness of the fault and the proximity of the top of the fault to the surface. Figure 6.16 shows the magnitude of the error, which even for the extreme case of the top of the fault close to the surface, the difference is a maximum of 5 percent. However, notice that the error tends to smooth the anomaly on the side over the basin and sharpen the anomaly on the shallow edge of the basin, giving a strong indication of which side is deeper.

As with the simple fault, a near-vertical dike can be modeled to evaluate depth from the surface and the effects of variation in dip. Gravity data observed over real dikes often suggest the existence of more complex structures, such as those that include multiple dikes. The anomalies can show multiple peaks and a broader scale, indicating that more of the dike exists at depth than at the surface. Figure 6.17 shows a gravity profile across one such dike swarm. Magnetic data were used to locate the major dikes in the swarm and gravity

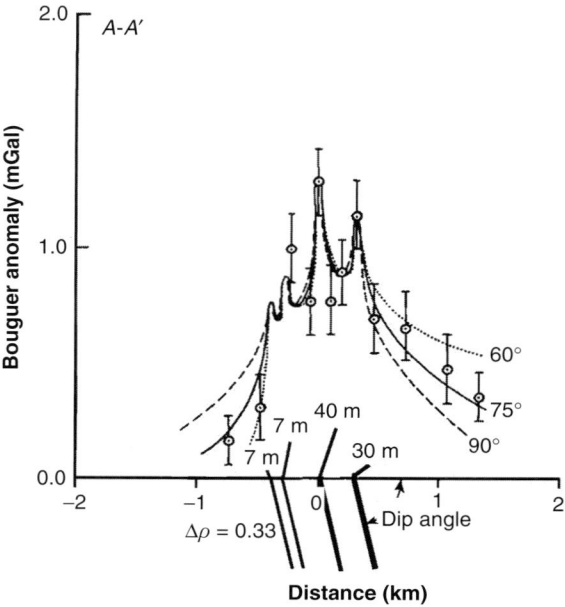

Figure 6.17 Example of the influence of dip on the gravity anomaly for a near-vertical dike (after Rothe and Long 1975).

models for dipping dikes were used to demonstrate the effect of dip on the anomaly. A dip of 75 degrees provides a reasonable fit to the data.

Other shapes can be utilized to evaluate two-dimensional gravity anomaly computations. Starting with the two-dimensional expression for a gravity anomaly

$$\Delta g = 2G\Delta\rho \iint \frac{\xi}{\xi^2 + \zeta^2} d\xi d\zeta. \tag{6.48}$$

In Eq. (6.48), we substitute the relation

$$\frac{\xi}{\sqrt{\xi^2 + \zeta^2}} = \sin\phi, \tag{6.49}$$

where

$$\frac{1}{\sqrt{\xi^2 + \zeta^2}} = \frac{1}{r} \tag{6.50}$$

and

$$d\xi d\zeta = r dr d\phi \tag{6.51}$$

represent a change to cylindrical coordinates, see Figure 6.18a. On breaking the two-dimensional area into radial zones and angles, and converting Eq. (6.48) to cylindrical coordinates, the relation becomes

$$\Delta g = 2G\Delta\rho \int_{\phi_i}^{\phi_{i+1}} \int_{r_k}^{r_{k+1}} \sin\phi dr d\phi \tag{6.52}$$

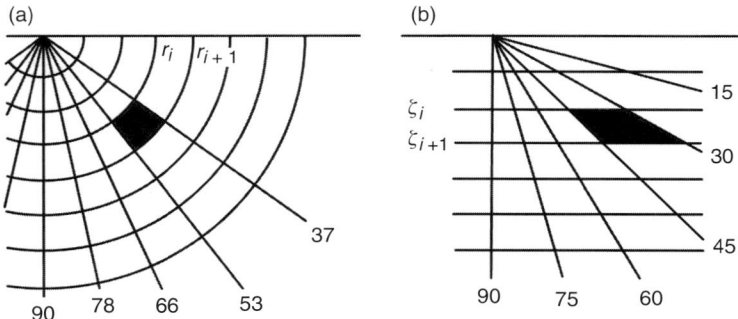

Figure 6.18 Other geometries useful for interpreting two-dimensional gravity anomalies. (a) Conversion to cylindrical coordinates. (b) Conversion to flat layers and radial lines from the origin.

and for a single zone

$$\Delta g = 2G \Delta \rho \, (r_{k+1} - r_k)(\cos \phi_i - \cos \phi_{i+1}). \tag{6.53}$$

Figure 6.18a shows the shapes that result when each compartment has the same value contribution to the anomaly. By choosing divisions that are equal units of radius and by dividing the angle into equal values of the cosine of the angle, each element will have the same value. Note that as the distance increases, the size of the blocks also increases. Density anomalies at greater distances have a reduced effect on the computed anomaly. The angles appropriate for equal values in each compartment are given by

$$\phi_n = \cos^{-1} \frac{n}{N}. \tag{6.54}$$

Another computational scheme can be devised by recognizing that $r \sin \phi = d\zeta$, so that the integration is trivial

$$\Delta g = 2G \Delta \rho \iint d\zeta \, d\phi = 2G \Delta \rho \, (\zeta_{k+1} - \zeta_k)(\phi_{i+1} - \phi_i). \tag{6.55}$$

In Eq. (6.55), polygons defined by equal thicknesses of depth and intersected by equal increments of the angle in Figure 6.18 (b) will produce polygons with equal contribution to the gravity anomaly.

6.12 Polygons of attracting mass in a horizontal sheet

The attraction of a density anomaly in the shape of a polygonal section of a flat and very thin plate is convenient for some geologic structures. By stacking such horizontal polygons one on top of another a massive body of arbitrary shape can be modeled; although, coherent structures like salt domes or granite plugs are the best shapes for this model. The gravity anomaly can be computed by breaking the structure into a stack of flat layers, where each

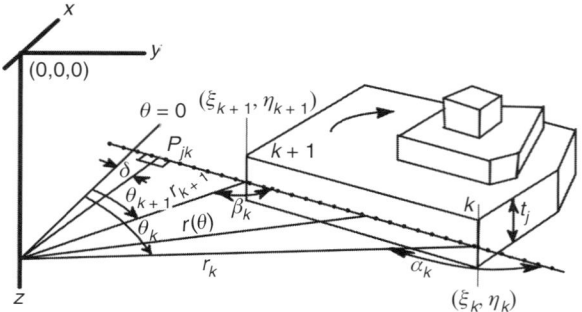

Figure 6.19 Reference geometry for a horizontal sheet model for gravity anomaly computation.

layer is a horizontal cross section of the structure at a particular depth. As with all sheet approximations, the thickness of the sheet should be less than half the depth to the sheet. In order to be consistent with the density for coherent structures the thicknesses of the sheets should provide a continuous model. However, the density and thickness are, by the non-uniqueness of potential data, coupled together and the important quantity is the anomalous mass per unit area of the sheet.

The derivation starts with the general expression for the vertical attraction of an infinitesimal element of mass that we integrate over the volume of the disturbing or anomalous mass as shown in Figure 6.19. The significance of the thin-sheet approximation is that the mean value theorem can be used to evaluate the integral over the thickness of the sheet. For simplicity the density may be considered a constant. Where density is variable, modeling would require separate polygons over areas of constant density, or numerical integration. Starting with Eq. (1.21) the basic equation for attraction is

$$\Delta g = \frac{\partial V}{\partial z} = -G \iiint \frac{\Delta \rho \, (\xi, \eta, \zeta) \, (z - \zeta) d\xi d\eta d\zeta}{\left\{ (x - \xi)^2 + (y - \eta)^2 + (z - \zeta)^2 \right\}^{3/2}}. \tag{6.56}$$

Integration over depth is computed first using the mean value theorem. The total gravitational attraction is given by the summation over M layers at mean depth for the jth layer of z_j and a thickness for the jth layer of t_j. Without loss of generality the values at the origin of the coordinate system on the surface can be set to $x = 0$, $y = 0$, and $z = 0$. Also, as in the derivation of Talwani and Ewing (1960), the equation can be transformed to cylindrical coordinates as shown in Figure 6.19. The integration is computed clockwise around the N_j corners of the polygon sequentially from $k = 1$ to N_j

$$\Delta g = G \Delta \rho \sum_{j=1}^{M} t_j \sum_{k=1}^{Nj} \int_{\theta_{jk}}^{\theta_{j(k+1)}} \int_{r=0}^{r=r(\theta_{jk})} \frac{z_j}{\left\{ r^2 + z_j^2 \right\}^{3/2}} r dr d\theta. \tag{6.57}$$

Values at other points are found by shifting the positions of the corners of the polygon. The area of integration is the triangle defined by the origin and the side of the polygon, the shaded area in Figure 6.19. By going around the polygon clockwise, the area on the near side is subtracted from the areas on the far side to leave only the area within the polygon. The integration with respect to radius gives

$$\Delta g = G \Delta \rho \sum_{j=1}^{M} t_j \sum_{k=1}^{Nj} \int_{\theta_{jk}}^{\theta_{j(k+1)}} \frac{z_j}{\{r^2 + z_j^2\}^{1/2}} \Bigg|_{r=0}^{r(\theta_{jk})} d\theta \tag{6.58}$$

$$\Delta g = G \Delta \rho \sum_{j=1}^{M} t_j \sum_{k=1}^{Nj} \left[\int_{\theta_{jk}}^{\theta_{j(k+1)}} d\theta - \int_{\theta_{jk}}^{\theta_{j(k+1)}} \frac{z_j}{\{r(\theta_{jk})^2 + z_j^2\}^{1/2}} d\theta \right]. \tag{6.59}$$

In the case where the origin is within the polygon, the first integral over θ goes to 2π and approaches the Bouguer Plate equation. If $r(\theta_{jk})$ is large relative to z_j, the second integral becomes small, as would be expected for a polygon that approaches a flat infinite plate. An expression for $r(\theta_{jk})$ can be found from Figure 6.19 by drawing a perpendicular, p_{jk}, to the kth face of the polygon at the jth depth. Then, from the relations among the angles the radius can be expressed as

$$r(\theta_{jk}) = \frac{p_{jk}}{\cos\left(\theta - \theta_{j(k+1)} + \beta_{jk} - 90\right)} = \frac{p_{jk}}{\cos\left(\theta - \theta_{jk} + \alpha_{jk} - 90\right)} \tag{6.60}$$

$$r(\theta_{jk}) = \frac{p_{jk}}{\sin\left(\theta - \theta_{j(k+1)} + \beta_{jk}\right)} = \frac{p_{jk}}{\sin\left(\theta - \theta_{jk} + \alpha_{jk}\right)}. \tag{6.61}$$

From these it can be seen that at the limit $\theta = \theta_{jk}$ the argument of the sin is α_{jk}, and at $\theta = \theta_{j(k+1)}$ the argument is β_{jk}. Integration of the first component gives the angle subtended by the side of the polygon. For the second component of integration the expression for $r(\theta_{jk})$ is inserted into Eq. (6.61) to get

$$\Delta g = G \Delta \rho \sum_{j=1}^{M} t_j \sum_{k=1}^{Nj} \left[(\theta_{j(k+1)} - \theta_{jk}) - \int_{\theta_{jk}}^{\theta_{j(k+1)}} \frac{z_j}{\left\{ \left(\frac{p_{jk}}{\sin\left(\theta - \theta_{j(k+1)} + \beta_{jk}\right)} \right)^2 + z_j^2 \right\}^{1/2}} d\theta \right]. \tag{6.62}$$

The second integral can be put into the form

$$\int_{\theta_{jk}}^{\theta_{j(k+1)}} \frac{1}{\left\{ 1 + \frac{z_j^2 \cos^2 \theta - \theta_{j(k+1)} + \beta_{jk}}{(p_{jk}^2 + z_j^2)} \right\}^{1/2}} d \left(\frac{z_j \cos\left(\theta - \theta_{j(k+1)} + \beta_{jk}\right)}{(p_{jk}^2 + z_j^2)^{1/2}} \right), \tag{6.63}$$

which integrates to

$$\Delta g = G\Delta\rho \sum_{j=1}^{M} t_j \sum_{k=1}^{Nj} \left[(\theta_{j(k+1)} - \theta_{jk}) - \sin^{-1} \frac{z_j \cos(\theta - \theta_{j(k+1)} + \beta_{jk})}{(p_{jk}^2 + z_j^2)^{1/2}} \Bigg|_{\theta_{jk}}^{\theta_{j(k+1)}} \right].$$

(6.64)

On evaluation at these limits and inserting the values as noted above, the final equation is

$$\Delta g = G\Delta\rho \sum_{j=1}^{M} t_j \sum_{k=1}^{Nj} \left[(\theta_{j(k+1)} - \theta_{jk}) - \sin^{-1} \frac{z_j \cos(\beta_{jk})}{(p_{jk}^2 + z_j^2)^{1/2}} + \sin^{-1} \frac{z_j \cos(\alpha_{jk})}{(p_{jk}^2 + z_j^2)^{1/2}} \right].$$

(6.65)

In rectangular coordinates, the equation for the line is $\xi = a_{kj}\eta + b_{kj}$, where the a_{kj} and b_{kj} are determined from the coordinates of the k and $k + 1$ corners of the polygon in the jth layer. For computation, generally the corners are given in their rectangular coordinates, (ξ_k, η_k) and (ξ_{k+1}, η_{k+1}). The components for Eq. (6.65) can be computed from the relations

$$p_{jk} = r_{k+1} \sin(180 - \beta_{jk}) = r_k \sin(180 - \alpha_{jk}) \tag{6.66}$$

$$r_k = (\xi_k^2 + \eta_k^2)^{1/2} \tag{6.67}$$

$$r_{k+1} = (\xi_{k+1}^2 + \eta_{k+1}^2)^{1/2} \tag{6.68}$$

$$r_{k,k+1} = ((\xi_k - \xi_{k+1})^2 + (\eta_k - \eta_{k+1})^2)^{1/2} \tag{6.69}$$

$$\cos\beta_{jk} = \frac{r_{k+1}^2 + r_{k,k+1}^2 - r_k^2}{2r_{k+1}r_{k,k+1}} \tag{6.70}$$

$$\cos(180 - \alpha_{jk}) = \frac{r_k^2 + r_{k,k+1}^2 - r_{k+1}^2}{2r_k r_{k,k+1}}. \tag{6.71}$$

The inversion of gravity data

7.1 Introduction

The inverse problem in potential theory is an exercise in reducing an infinite distribution of possible solutions down to one that is physically realistic and acceptable to an experienced interpreter. This chapter formalizes the determination of distributions of densities given assumptions concerning the properties of the density anomalies and the causative structures. Although the densities of materials surrounding a gravity observation uniquely determine the gravity field, the gravity data alone cannot determine a unique distribution of densities. The density can vary smoothly within a geologic structure, or it can be discontinuous, such as at boundaries of distinct geologic units. In either smooth or discontinuous density distributions there exists an infinite variety in the densities that can be found to fit any set of gravity observations. The only restriction is that the anomalous densities fall above some maximum depth. In general, density is neither a smooth nor uniform function of position. Density is discontinuous across boundaries at all scales, from mineral grain boundaries to the edges of major geologic structures. The details of the density structure are usually so complex that a complete definition would be neither practical nor advisable. In most cases, average or bulk densities for geologic units may be constrained to fall within narrow limits and these limits may be used to limit the density models that generate the gravity anomaly. In effect, the limitations placed on density by the composition and physical condition of geologic units to a large extent define acceptable density structures. On the other hand, the non-uniqueness of gravity data cannot be escaped. There exists an infinitely large set of density models that will satisfy a given gravity anomaly and most of these are not physically realistic. The task of gravity data inversion is to automatically find geologically acceptable density structures that also satisfy the observed gravity anomalies. Hence, reasonable densities for the lithology and acceptable shapes for structures are used to constrain the search for the density structure in gravity data inversion. In the inverse problem, the assumptions are formulated as constraint equations, sometimes referred to as the cost functions. The role of the cost functions or constraints is to add sufficient information to allow a unique solution for a density model consistent with the observed anomalies and expressed constraints.

In the general case, the inverse problem is defined by observed data, d, and model parameters, m. The observed data are usually the observed gravity values, Δg, although other observations such as seismic velocity, observed geologic structures, and measured rock density may also serve as additional data. The model parameters are the densities and their spatial distribution. The inverse problem is to find density as a function of position, $\Delta\rho(x, y, z)$, where the model parameters are density values at each location. More often the inverse problem is formulated by using constant density values within defined structures, where the model parameters represent some component of the structure such as depth, thickness, or width.

For any observed gravity anomaly there is a function or set of functions, F, relating the data and the model parameters such that

$$0 = F(d, m). \tag{7.1}$$

In Eq. (7.1) the objective is to solve for the model parameters, m, given the data, d. This is usually accomplished by attempting to make Eq. (7.1) a linear relation, such that the function is expressed in terms of constant multipliers of the data and model parameters. Because the relation may contain a number of linear relations, Eq. (7.1) takes an implicit form

$$0 = \sum_{j=1}^{n} F_{ij}d_j + \sum_{j=n+1}^{n+m} F_{ij}m_{j-n} = F_{ij} \begin{vmatrix} d_j \\ m_{j-n} \end{vmatrix}, \tag{7.2}$$

where there are n data values and m model parameters. In the more common formulations, the first term in Eq. (7.2) is an identity and $F = |-I, G|$ so that the equation can be expressed more simply as an explicit equation

$$0 = G(m) - d \tag{7.3}$$

or in a strictly linear form

$$0 = \sum_{j=1}^{m} G_{ij}m_j - d_i. \tag{7.4}$$

In the explicit form there is no guarantee that model parameters can be found that make Eq. (7.4) identically zero. The solution gives an estimate of the model parameters, m^{est}. Equation (7.4) represents an error when the estimated model parameters are substituted for the unknown model parameters. In the least square error solution, the cost function that is minimized in order to find the solution for the model parameters is the length or magnitude of the square of the error, computed from

$$e_i^T e_i = \left(\sum_{j=1}^{m} G_{ij}m_j^{est} - d_i \right)^T \left(\sum_{j=1}^{m} G_{ij}m_j^{est} - d_i \right). \tag{7.5}$$

Unless carefully designed, the least squares solution may be unstable and yield widely varying and unrealistic density anomalies. As an alternative cost function, the length or magnitude of the model parameters can be minimized, leading to a minimum length solution. As the name implies, the minimum length solution suppresses variations in the density anomalies. A combination of these two cost functions gives a generalized inverse solution, where a choice has to be made concerning how much smoothing to permit from the contribution of the minimum length solution versus the possibility of obtaining a more accurate solution by finding the solution with the least square error. Ultimately, the goal of data inversion should be to find the solution with the greatest likelihood of being a correct representation of the observed data and constraining conditions. The generalize inverse when the data and model parameters are weighted by their uncertainty does give the maximum likelihood solution, but only if the errors are random and satisfy a Gaussian distribution.

7.2 Depth of a basin as layer thicknesses

In previous chapters we have considered the direct problem, the computation of the gravity field by structures of specific shapes, such as a sedimentary basin or horizontal polygonal plates. In this chapter, we formulate the inverse problem, the use of gravity data to automatically solve for parameters which define the shape of the density model. The non-uniqueness of this inversion is circumvented by limiting the allowable shapes of a density model or in some way constraining the density model. As a first example, we will examine the relation between the thickness of sediments in a sedimentary basin and the gravity anomaly as shown in Figure 7.1. The gravity anomalies are to be used to extrapolate depths to other parts of the basin.

Many larger sedimentary basins, such as the Delaware basin in southeastern New Mexico or the Michigan basin, take the general shape of a shallow dish. In such basins the depth of the basin is shallow relative to the width of the basin, an important factor in formulating an inverse model. At any given point, the gravity anomalies of the lower-density sediments may be approximated by the gravity anomaly generated by a Bouguer plate of constant density and thickness proportional to the depth. For this simple model, the gravity anomalies may be approximated in Eq. (7.4) by a linear function of the thickness according to the equation

$$\Delta g_i - a_0 = a_1 h_i. \tag{7.6}$$

The unknown, a_0, represents regional anomalies or shift in the datum plane used to compute the anomaly and is independent of the structure of interest. The unknown, a_1, is equivalent

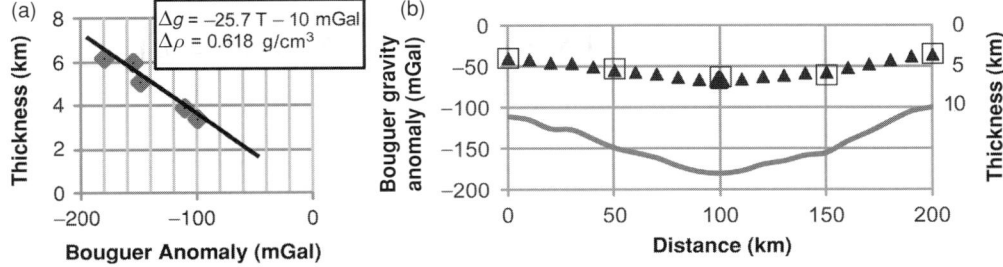

Figure 7.1 Depth estimates for the Delaware basin, of southeast New Mexico. (a) Relation between depth from drilling or seismic data and observed Bouguer anomaly. (b) Calibration depths are the open squares. The solid line corresponds to the observed Bouguer anomaly and the triangles are the depths computed from the Bouguer anomalies.

to the term $2\pi G\Delta\rho$ in the Bouguer plate correction and is linearly proportional to the density contrast. Equation (7.6) may be solved for the coefficients a_0 and a_1, if independent pairs of values for Δg and h are available, perhaps from two wells that penetrate to two different basin depths at two separate locations, or, perhaps, from one well and a value at the edge of the basin where h is negligibly small. Figure 7.1 shows data taken from Adams and Keller (1996) from the Delaware Basin of New Mexico. A complicating factor in using gravity anomalies to solve for basin depth for most basins is the tendency for the average density of the sediments to increase with depth, decreasing the density contrast as a function of basin depth. Equation (7.6) may be extended to include the effects of increased density by extending the power series

$$\Delta g = a_0 + a_1 h + a_2 h^2 + \cdots + a_n h^n + \cdots. \tag{7.7}$$

The solution for the coefficients, a_i, could be found by least squares regression, provided that the scatter in the data can be defined by the values of depths available from well data. Data for the Delaware Basin in Figure 7.1a follow a nice linear relation between depth to basement and Bouguer anomaly. A density contrast for a least squares fit to the data suggest a density contrast of 0.618 g/cm^3. Typically, the number of significant coefficients is limited to three because the fourth, a_3, requires resolution of changes in the curvature of the relation between Δg and h that are typically hidden in the noise of the data.

Equation (7.7) solves the direct problem of estimating gravity anomalies from basin depths and provides estimates consistent with observed depths. The inverse of Eq. (7.7) can be obtained by direct solution or, for higher orders, by successive approximation. As an alternative, we could pose the problem initially in the form

$$h = b_0 + b_1 \Delta g + b_2 \Delta g^2 + b_3 \Delta g^3 + \cdots \tag{7.8}$$

and solve for the coefficients b_i, by a least squares regression. Equation (7.8) solves for the depth of the basin at any point given that the gravity anomaly above that point is known. Again, for a simple basin, the depth is proportional to the ratio of the gravity anomaly to the density contrast,

$$h = b_0 + \frac{\Delta g}{2\pi G\rho} + b_2 \Delta g^2 + \cdots. \tag{7.9}$$

Figure 7.1b shows the gravity anomalies and the interpreted thickness of the sediments. The depth is known at a few points, and these depths can be extrapolated away from the well locations by using the solution for the coefficients. The most significant term is the b_1 term, which corresponds to the Bouguer plate first approximation. The b_2 term corrects the relation for variation in the density with depth. Density typically increases with increased depth, but this term could also compensate for variations in rock type. Higher-order terms are seldom significant, except where there exist a significant vertical variation in rock type that is consistent throughout the basin. As might be expected, this interpretation is not unique. Adams and Keller (1996) considered a much lower density contrast for the sediments. Consequently, their models required significant decreases in the density of the crust below the basin.

7.3 Depth in a basin as an overdetermined inverse problem

In many basins, particularly some of the narrow Triassic rift basins, the thickness of the basin can vary sufficiently rapidly within a distance comparable to the depth. In these cases the accuracy of using the Bouguer plate anomaly or linear regression to find depth is questionable. Generally, if the estimated depth varies more than 10 percent within a distance equivalent to the estimated depth of the basin, then a more accurate two-dimensional model should be considered. When the depth of the basin changes rapidly with distance, the variation in depth of adjacent areas of the basin will significantly affect the gravity anomaly.

The gravity value is determined by contributions from all the adjacent depths in the basin. For such a model, a basin may be divided into vertical columns or for elongated basins into vertical dikes. The exact equations for the density model will depend on the shape of the vertical column and the precision needed in the computation. For simple versions of this type of inversion, the structure can be assumed to be composed of a material of uniform density. Hence, the density is constant and the equations are set up to find some parameter in the dimensions of the structure. The relation between observed gravity and the model parameters is defined in Eq. (7.3) The gravity values, Δg_i, are functionally related to the model parameters through some function $G(m)$. If the expression for the function, $G(m)$, is linear, it would depend on the position of the ith gravity point relative to the jth column. A linear equation for the attraction at the ith point can be written as

$$\Delta g_i = \sum_{j=1}^{m} G_{ij} m_j, \qquad (7.10)$$

where m is the total number of columns considered in the inverse problem and is significantly less than the number of independent gravity observations. In the overdetermined inverse problem, a simple model with typically only a few unknown parameters is found that best fits a greater number of independent gravity observations.

The values of the G_{ij} are functions of shape (e.g. thickness h) and density contrast. In general, we can choose to solve for density or shape, but the non-uniqueness of potential data prevents solving for both shape and density without adding more constraints. If we choose to fix the shape and solve for a density anomaly, $\Delta \rho_j$ for the jth column, the contribution from each column can be expressed as a product $G_{ij} \Delta \rho_j$, where the densities are the model parameters. In this form the equations for the influence of the basin anomaly on Δg_i are linear and take the form

$$\Delta g_i = \sum_{j=1}^{m} G_{ij} \Delta \rho_j. \qquad (7.11)$$

A solution is generally possible provided that the number of gravity values, i, is greater than the number of density contrasts, m. If, as in the basin model above, the densities vary only slightly and the major contributions to the gravity field are from variations in the basin depth, h, the function, G in general cannot be linearly related to depth. The non-linear equation can be solved iteratively by Newton's method

$$\Delta g_i = G_i(h^{est}) + \sum_{j=1}^{m} \frac{\partial G_i(h^{est})}{\partial h_j} \Delta h_j + \cdots, \tag{7.12}$$

where h^{est} represent the current guess for the depths, perhaps as first computed from the simple Bouguer plate approximation, and the Δh_j are corrections to the current best guess. The values of Δh_j are computed using updated basin depths until they converge to a solution. Equation (7.12) can be written in the form of Eq. (7.11) if we identify the gravity anomalies as the gravity anomalies minus the estimated gravity values, the first term in Eq. (7.12). The Jacobian matrix is the partial derivatives in the summation in Eq. (7.12). There are multiple physical models that can be employed in this type of inversion. As an example, the cross section of a long narrow basin, very much like a typical Triassic basin along the east coast of North America could be modeled using two-dimensional vertical sheets. The equation for a vertical sheet is

$$\Delta g(x_i) = \sum_{j=1}^{m} 2G \Delta \rho t_j \ln \left(\frac{(x_i - x_j)^2 + z_1^2}{(x_i - x_j)^2 + z_j^2} \right), \tag{7.13}$$

where t_j is the thickness of the sheet, $x_i - x_j$ is the horizontal distance to the sheet, z_j is the depth of the bottom of the sheet, and z_1 is the depth to the top. The derivative of Eq. (7.13) with respect to z_j, which would be used in the Jacobian matrix of Eq. (7.12), is

$$\frac{\partial \Delta g(x_i)}{\partial z_j} = -4G \Delta \rho t_j \frac{z_j}{(x_j - x_i)^2 + z_j^2}. \tag{7.14}$$

Figure 7.2 is a test of the method of Eq. (7.12) using a model based on Eqs. (7.13) and (7.14). The gravity anomalies are generated using vertical sheets with a separation of 1.25 km extending from a depth of 1.0 km (approximate depth of overlying sediments) to the depth on the basin. Figure 7.2a shows the raw data and Figure 7.2b (solid line) is the starting model based on the first approximation by a Bouguer plate anomaly. The density contrast chosen was 0.05 g/cm^3. A more correct density for this basin would be 0.09 g/cm^3, but the lower density better illustrates the effects of the inversion. The computed depth to the bottom of the basin is the dotted line. In the inversion, damped least squares were used and the solution was constrained to be at or below the base of the overlying sediments. Also, a regional trend was removed. In most models of this kind, uncertainty at the edges where the model does not represent sources outside the model can lead to instabilities. The inversion for basin depth, the dotted line in Figure 7.2b, shows steepening in the area of the edge fault of the basin. Using a lower density would eventually lead to instabilities in this type of inversion. Inversions of this type are proxies for downward continuation, where the gravity anomaly is continued down to the depth of the basin. In downward continuation, an attempt to exceed the depth of the source of the anomaly can generate instability, particularly where an equivalent to a point source exists at that depth.

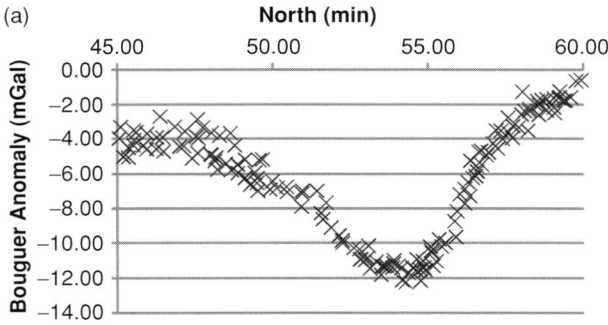

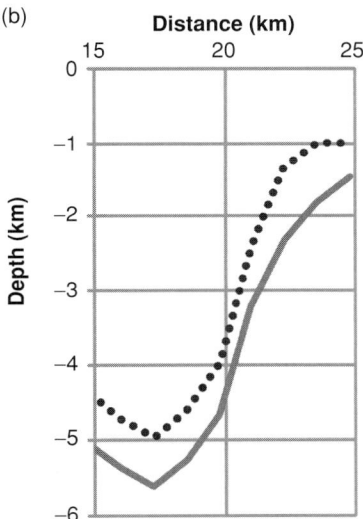

Figure 7.2   Inversion for the structure and depth of a Triassic Basin. (a) The gravity data used in the reduction. (b) Solution (dotted line) for depth using solid line as starting model.

All the iterations require the inversion of the Jacobian matrix. The solution is non-linear because the derivatives change with each correction to the model. A solution using this method also requires many more gravity values than unknown depths in the basin. Convergence can be aided by using damped least squares, which is introduced by adding a constant to the diagonal of the Jacobian matrix. This will avoid the situation where early iterations push the depth too deep and introduce instabilities. Typically, 5 to 10 iterations are needed to converge to a reasonable solution, one that does not change significantly with further iterations. An alternative technique for inversion of the matrix would be singular value decomposition, which by dropping small eigenvalues effectively removes the unstable components of the inversion.

7.4 Stripping by layers

Anomalies are often associated with particular depth zones that can be stripped from the total measured gravity. For example, to model the Earth's crust in areas where the lower crust may not contribute significantly to the observed anomaly, the observed gravity anomalies could be separated into those associated with changes in crustal thickness, the Moho depth, and those associated with density anomalies in the shallow crust. Stripping works best when the sources of anomalies can be associated with distinct and separated depth zones and when the source of the gravity anomalies can be associated with short-wavelength structures. The total anomaly is the sum of the anomalies from the separated depths

$$\Delta g(x, y) = \Delta g_1(x, y) + \Delta g_2(x, y) + \cdots, \tag{7.15}$$

where the first could be caused by the shallow structures and the second by the deeper structures.

The first step in stripping is to separate the short wavelengths from the longer wavelengths. The shorter wavelengths will be associated with density anomalies that must be in the shallow layer and the longer wavelengths with density structures that could be in either the deeper or the shallow layer. This can be accomplished by using many different smoothing filters, but the most logical is upward continuation. The upward continued field is subtracted from the original gravity field. The difference represents near-surface anomalies so it is appropriate that the upward continuation distance is equal to the depth to the top of the second layer. The difference between the upward continued anomalies and the observed anomalies consists only of those components of the original anomaly that must originate from depths at or above the continuation distance. Remember, because of the non-uniqueness of potential data, shallow structures containing longer wavelengths are included in those that could occur in the lower layer. The anomalies that must be associated with the surface layer on the basis of their short wavelength are then found in the residual

$$\Delta g_1(x, y, 0) = \Delta g(x, y, 0) - \Delta g(x, y, h_1), \tag{7.16}$$

where h_1 is the continuation distance, or depth to the top of the second layer. The observed anomaly has thus been divided into anomalies that must be associated with structures above the second layer and those anomalies that could be in or below the second layer. The surface anomaly could then be continued to a depth defined by a third layer in order to identify those anomalies that must be above that depth but below the shallow layer. The anomalies that must be above the second layer, and do not include the anomalies of the first layer are given by

$$\Delta g_2(x, y, 0) = (\Delta g(x, y, 0) - \Delta g_1(x, y, h_1)) - \Delta g(x, y, h_2)$$
$$= \Delta g_1(x, y, h_1) - \Delta g(x, y, h_2), \tag{7.17}$$

where h_2 is the continuation distance to the top of the third layer. A significant ambiguity and problem with stripping is that long-wavelength anomalies generated by shallow structures

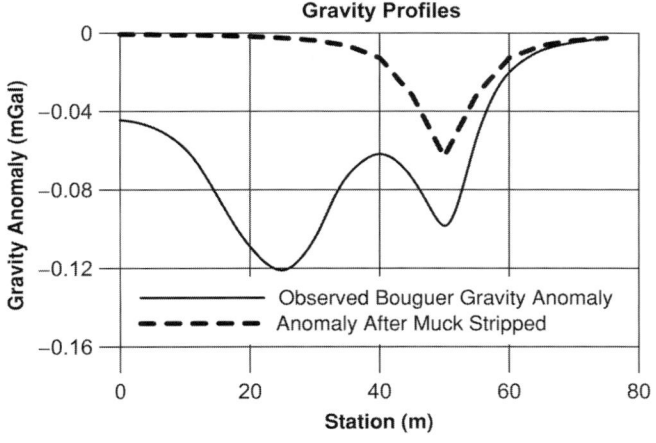

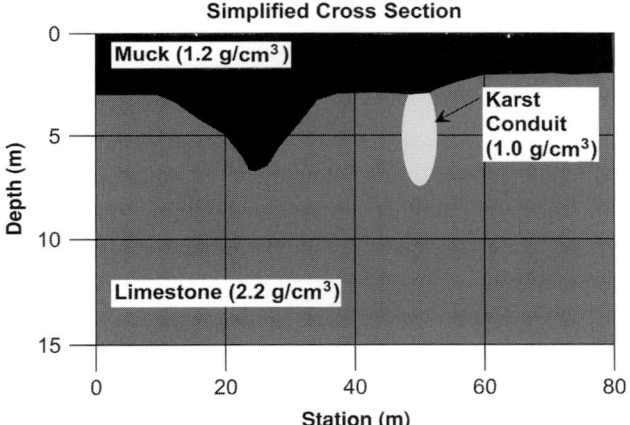

Figure 7.3 Example of stripping in inversion by combining surface geology with the inversion process.

may be pushed to unrealistic depths. Hence, stripping is appropriate only when the source anomalies are limited in their lateral extent to wavelengths less than the depth to the layer.

Stripping may be made possible by other types of geophysical data. For example, the locations of cavities and dissolution conduits are often the main target of interest for microgravity surveys in karst terrain. The objective for these microgravity surveys is to locate cavities on the basis of their negative anomalies. However, the shallow near-surface weathering layer over limestone can vary in thickness and its density contrast with the limestone can generate anomalies that dominate the gravity anomalies and mask the anomaly from a cavity. Figure 7.3 shows a simplified cross section in karst terrain. The overlying weathered layer and soil has a thickness of 2 to 7 m and an observed Bouguer anomaly of between 0 and −0.12 mGal. Using a thickness profile for the overlying layer obtained from a series of shallow auger holes and a density obtained from field samples, the influence of the surface layer is removed. Having stripped the near-surface layer, the remaining anomalies correspond to those from density structure in the limestone, such as a water-filled conduit.

7.5 Formulation of an underdetermined inverse problem

Gravity anomalies such as the basin anomaly illustrated in Figure 7.2 represent contributions from a three-dimensional distribution of density anomalies distributed throughout the basin and the underlying crust. Without loss of generality, we can assume that Figure 7.2 represents the cross section of a long narrow basin, and the gravity anomaly is the expression of a two-dimensional distribution of density anomalies. A model can be defined by partitioning the cross section into conveniently shaped segments, for example, rectangular rods for which the attraction can be computed. For small compact polygonal rods, the equation for a horizontal rod is usually sufficient for the precision of gravity anomalies. Similarly for a three-dimensional model, a three-dimensional distribution of small spheres could be used to approximate the attraction of a smaller and more complex three-dimensional shapes for which more complex equations would be required. Figure 7.2 showed a model for a two-dimensional basin. The model consists of many horizontally layered polygons of unknown density contrast. The gravity anomaly in this example is sampled by a significantly smaller number of gravity observation points. The gravity anomaly at each observation point contains contributions from each density anomaly. The magnitude of the contribution is proportional to the density contrast and inversely proportional to their separation. However, the significant point here is that the number of unknown density contrasts can far exceed the number of gravity observations.

For any model, then, the influence of the jth block on the ith gravity anomaly can be designated as in Eq. (7.11) by $G_{ij}\Delta\rho_j$. The ith gravity anomaly is the cumulative effect of all the density anomalies, and can be written in the form

$$\Delta g_i = \sum_{j=1}^{m} G_{ij}\Delta\rho_j. \tag{7.18}$$

Because there are many more density values than there are data points the inverse problem is underdetermined in addition to being non-unique. There exists an infinite number of solutions and the least square error is no longer a useful constraint to use to find a solution. Hence, a different approach is needed. Instead of minimizing the error between the observed data and theoretical anomaly computed from the density model, the length of the density anomalies is minimized. This is referred to as the minimum length solution.

7.6 The minimum length solution of the underdetermined inverse problem

In the underdetermined gravity inverse problem, there are many more density values to be computed than there are observations of the gravity anomaly. Furthermore, because the observed gravity depends on all the density anomalies, an infinite number of solutions exist that can exactly fit the observed data. This is a classical manifestation of the non-uniqueness of potential data. In order to choose one of these solutions, additional information in the form of a priori data must be supplied. A priori data are constraints that are not given in the equation relating gravity anomalies to density structure. Such information reduces the range of solutions or even makes the solution unique. In effect, these constraints permit the introduction of the interpreter's bias toward a desired solution.

The simplest a priori information is the constraint that the magnitude of the density anomaly, its length, be minimized. One wishes to find a solution for density structure that fits the gravity anomalies exactly and satisfies the condition that the sum of the squares of the density anomalies is as small as possible. Hence, we minimize the length defined by

$$\Delta\rho^T I \Delta\rho = \sum_{j=1}^{m} \Delta\rho_j^2, \tag{7.19}$$

where I is the identity matrix, under the condition from Eq. (7.18) that

$$\Delta g = G\Delta\rho. \tag{7.20}$$

The condition that the length is minimized is equivalent to the set of constraint equations defined by setting each density anomaly to zero. To develop the minimum length solution, the Lagrange multipliers λ_j are used. The function to be minimized is

$$\Phi(\Delta\rho) = \sum_{j=1}^{m} \Delta\rho_j^2 + \sum_{i=1}^{n} \lambda_i \left[\Delta g_i - \sum_{j=1}^{m} G_{ij}\Delta\rho_j \right], \tag{7.21}$$

where there is one Lagrange multiplier for each measurement error. Differentiation of Eq. (7.21) for each density and setting the derivative to zero gives a set of equations

$$\frac{\partial\Phi(\Delta\rho)}{\partial\Delta\rho_q} = 0 = \sum_{i=1}^{N} 2\Delta\rho_i \frac{\partial\Delta\rho_i}{\partial\Delta\rho_q} - \sum_{i=1}^{n} \lambda_i \sum_{j=1}^{m} G_{ij}\frac{\partial\Delta\rho_j}{\partial\Delta\rho_q} \tag{7.22}$$

$$0 = 2\Delta\rho_q - \sum_{i=1}^{n} \lambda_i G_{iq}, \tag{7.23}$$

which can be written in matrix form as

$$\Delta\rho = G^T \frac{\lambda}{2} \tag{7.24}$$

and substituted into Eq. (7.20) to solve for the λ

$$\lambda = 2\left(GG^T\right)^{-1}\Delta g. \tag{7.25}$$

We then substitute Eq. (7.25) into (7.24) to get the minimum length solution

$$\Delta\rho^{est} = G^T\left(GG^T\right)^{-1}\Delta g. \tag{7.26}$$

The minimum length solutions have interesting consequences that can be examined with simple models as illustrated in Figure 7.4. In Figure 7.4a the true density was a unit density contrast in one block in the top layer. The minimum length solution reduces the value in the one block down to 0.92, and uses negative values in the top layer balanced by positive values

(a)	Shallow model							
				1.00				

Minimum length solution							
−0.02	−0.04	−0.07	0.92	−0.07	−0.04	−0.02	−0.01
0.01	0.03	0.11	0.21	0.11	0.03	0.01	0.00

(b)	Deep model							
					1.00			

Minimum length solution							
0.00	0.01	0.03	0.11	0.21	0.11	0.03	0.01
0.01	0.02	0.04	0.07	0.09	0.07	0.04	0.02

Figure 7.4 Minimum length solutions to the gravity anomaly from two simple models.

in the lower layer to provide a minimum length solution. In the minimum length criteria, the larger values have significantly greater influence and are suppressed. The smaller density values have minimal influence. Overall, the minimum length solution is reasonable, but not exact, when the causative anomalies are in the shallow layer. Figure 7.4b shows the minimum length solution for the anomaly from a deeper source. A dominant portion of the anomalous density is shifted to the shallow layer, suggesting that most of the anomaly could have come from the shallow layer. In effect, the minimum length solution spreads the anomalous deep density structure over a range of depths above their actual position.

Gravity data inversion and downward continuation are linked (Fedi and Pilkington, 2012). The gravity anomalies from shallow sources cannot be downward continued deeper than the top layer. However, the gravity anomaly from a deep source can be downward continued to any depth at or above the deep layer. The differences between various inversion techniques are principally the criteria they use to determine how to distribute the density anomalies over their valid depth range.

There are many additional constraints that can be introduced. One useful constraint is to minimize the difference between adjacent density values. For this constraint the difference between two density values is set to zero such that $0 = \Delta\rho_i - \Delta\rho_{i+1}$ for all i. In matrix form these constraints form the flatness matrix, D, given by

$$
\begin{vmatrix}
1 & -1 & 0 & \cdots & 0 \\
0 & 1 & -1 & \cdots & 0 \\
0 & 0 & 1 & \cdots & 0 \\
\vdots & \vdots & \vdots & \ddots & \vdots \\
0 & 0 & 0 & \cdots & -1
\end{vmatrix}
\begin{vmatrix}
\Delta\rho_1 \\
\Delta\rho_2 \\
\Delta\rho_3 \\
\vdots \\
\Delta\rho_n
\end{vmatrix}
= D\Delta\rho. \tag{7.27}
$$

The length of these constraint equations from Eq. (7.19) is

$$
(D\Delta\rho)^T D\Delta\rho = \Delta\rho^T D^T D\Delta\rho = \Delta\rho^T W_m \Delta\rho, \tag{7.28}
$$

where the $W_m = D^T D$ is a weighing factor that influences the calculation of densities. Higher-order derivatives spread the weight over more values and give smoother results to the solution. Alternatively, the weighting factor could be used to place more or less emphasis on individual density values. Typically, the weighting matrix would be the inverse of the covariance matrix of the model parameters, which for simplicity is usually assumed to be a diagonal matrix with elements proportional to the inverse of the expected variance of the densities. Off-diagonal elements, as in the flattening matrix, suggest a correlation among density values.

In addition to flattening or weighting the solution, the values of density may be known a priori to be close to a value dictated by the rock type. In this case there is an a priori value m_a for the model parameters m and the minimum length solution is found by minimizing

$$(m - m_a)^T W_m (m - m_a).$$
(7.29)

The solution for a general weighting matrix with a priori information on the densities is

$$\Delta\rho = m_a + W_m^{-1} G^T \left(G W_m^{-1} G^T \right)^{-1} (\Delta g - G m_a).$$
(7.30)

If the weight matrix is the identity matrix and no a priori densities are provided, the solution reduces to the minimum length solution given in Eq. (7.26).

Many additional constraints, particularly those utilizing other geophysical techniques, will allow a stable least square error solution for at least some of the density values. For application to these mixed minimum length and least square error constraints, an appropriate solution would be to minimize some combination of the least square error and the length of the density solution. In addition, some gravity observations will be more accurate than others. In order to give more weight to the more precise gravity values, a weight function based on the inverse of the covariance matrix is generally used. The weight function applied to the least square error is

$$(\Delta g - G m)^T W_e (\Delta g - G m).$$
(7.31)

In most analyses, the diagonal of W_e is defined as the inverse of the variance of the gravity observations and off diagonal components are ignored. Ignoring off-diagonal components can lead to bias in the solution, particularly in the situation where a few observations are clustered tightly together. Also, some gravity values could be assigned an arbitrarily large variance to suppress their influence on the solution where they may be caused by structures of little interest in the inversion. Clustered observations would be highly correlated and using the off-diagonal components would properly weight these values as an average. The weighted least squares solution is

$$\Delta\rho = \left(G^T W_e G \right)^{-1} G^T W_e \Delta g.$$
(7.32)

The solution to the mixed minimum length and least squares problems is found by minimizing

$$(\Delta g - G m)^T W_e (\Delta g - G m) + \varepsilon^2 (m - m_a)^T W_m (m - m_a),$$
(7.33)

where ε^2 is a constant that determines the relative influence of the minimum length and least squares solutions. The choice of ε^2 is an art, a choice the interpreter has to make on the basis of the character of the inverse solution. There are two equivalent forms for the solution to the mixed problem

$$\Delta\rho = \Delta\rho_a + \left(G^T W_e G + \varepsilon^2 W_m\right)^{-1} G^T W_e \left(\Delta g - G\Delta\rho_a\right), \tag{7.34}$$

which reduces directly to the least squares solution with ε^2 and $\Delta\rho_a$ set to zero, and

$$\Delta\rho = \Delta\rho_a + W_m^{-1} G^T \left(G W_m^{-1} G^T + \varepsilon^2 W_e^{-1}\right)^{-1} \left(\Delta g - G\Delta\rho_a\right), \tag{7.35}$$

which similarly reduces directly to the minimum length solution with ε^2 and $\Delta\rho_a$ set to zero.

7.7 Seismic velocity as a constraint

Rather than finding constraints for the inversion of gravity data, gravity data may be used as a constraint for the inversion of other types of geophysical data. Kaufmann and Long (1996) used gravity data to constrain an inversion for seismic velocity. By ignoring the minor changes in average composition of the crust, Birch's (1961) relation may be differentiated to relate changes in velocity anomalies to changes in density anomalies as in Eq. (6.2)

$$\Delta V_p = b\Delta\rho, \tag{7.36}$$

where V_p is the compressional wave velocity and b is the slope. For shallow crustal rocks in the velocity range of 3.5 to 6.4 km/s the slope was assumed to be linear with a value of 6.15. Equation (7.36) relates the density anomaly to the velocity anomaly. The relation between the density anomaly and gravity anomaly was computed using the equations of Nagy (1966) for a rectangular prism. This model allows computing the gravity anomalies from a matrix relation

$$\Delta g_i = A_{ij}\Delta\rho_j. \tag{7.37}$$

Equation (7.37) is expressed in terms of the velocity anomaly by substituting for the density anomaly from Eq. (7.36) to get

$$\Delta g_i = A_{ij}\left(\frac{\Delta V_j}{b}\right). \tag{7.38}$$

The seismic velocity inversion component used conventional tomography, where the travel time anomaly was equated to the travel time in each block and its velocity perturbation. The form of the equation is

$$\Delta t_i = \sum_{j=1}^{N}\left(\frac{l_{ij}}{V_j}\right)\left(\frac{\Delta V_j}{V_j'}\right) = G_s\Delta V_j. \tag{7.39}$$

Combining Eqs. (7.37) and (7.39) gives in matrix form

$$\begin{bmatrix} \Delta g \\ \Delta t \end{bmatrix} = \begin{bmatrix} A \\ G \end{bmatrix}[\Delta V]. \tag{7.40}$$

The solution was found by using Eq. (7.34) without specifying a priori values because the absolute values for the velocities were removed in solving for the velocity perturbations. Figure 7.5a and Figure 7.5b shows the inferred velocity structure using, respectively, Eq. (7.38) for only seismic data, and Eq. (7.40) for a joint inversion of seismic and gravity data. Figure 7.5 was contoured from the original data of Kaufmann and Long (1996). The northwest half of the area was sparse in seismic data and consequently the inversion shows significant noise. The joint inversion greatly reduces the scatter in the area deficient in seismic coverage. The southeast half of the area where seismic data dominated were smoothed by the addition of gravity data, but not significantly changed. The most notable difference in the area of sparse seismic data was to remove the noise in the inversion for seismic data alone. The advantage of a joint inversion over an inversion for gravity or seismic data alone is that the combined data set facilitates extrapolation into areas where one or the other is inadequate. In particular, the gravity data facilitate extrapolating velocities into areas of sparse seismic coverage. Also, in areas of sufficient seismic coverage, the seismic data help to constrain the depth of the density anomalies.

7.8 Maximum likelihood inversion

The minimum length and the least square error approaches to the inversion of gravity data, with or without additional constraints, provide the classical approach to finding acceptable density distributions in the Earth. A typical approach to the minimum length and the least square error solutions will assume that the errors are uncorrelated Gaussian distributions. All of these methods explained so far in this chapter minimize deviations from specific properties that gravity data and density structure should satisfy. These methods use the results of many experiments and theory to define uncertainties and probability density functions. This is the philosophical approach of the "Frequentists" who assume that the solution parameters are fixed, but unknown constants. An alternative philosophical approach taken by the "Bayesians" is to consider parameters as random variables with probability density functions. They use the rules of probability to make inferences about parameters, a priori information, and observed data in order to examine a range of solutions. The solution to Eq. (7.1) that has the greatest probability of being correct is the maximum likelihood solution. Because this is the fundamental objective in gravity data inversion an examination of the maximum likelihood solution technique is appropriate, although rarely used.

Utilizing the "Frequentists" approach, the results of multiple observations are usually assumed to fit a traditional Gaussian error distribution where the errors can be assigned a standard deviation for assessing certainty in the solution. Menke (1984, Chapter 5) demonstrates that two conditions are required for these solutions to correspond to the maximum likelihood solutions. First, the probability density functions for the data must fit a Gaussian error function. Second, the weighting matrix must be the inverse of the full data covariance matrix. The full data covariance matrix includes off diagonal elements representing cross correlations of the data. In contrast, the full "Bayesian" approach allows

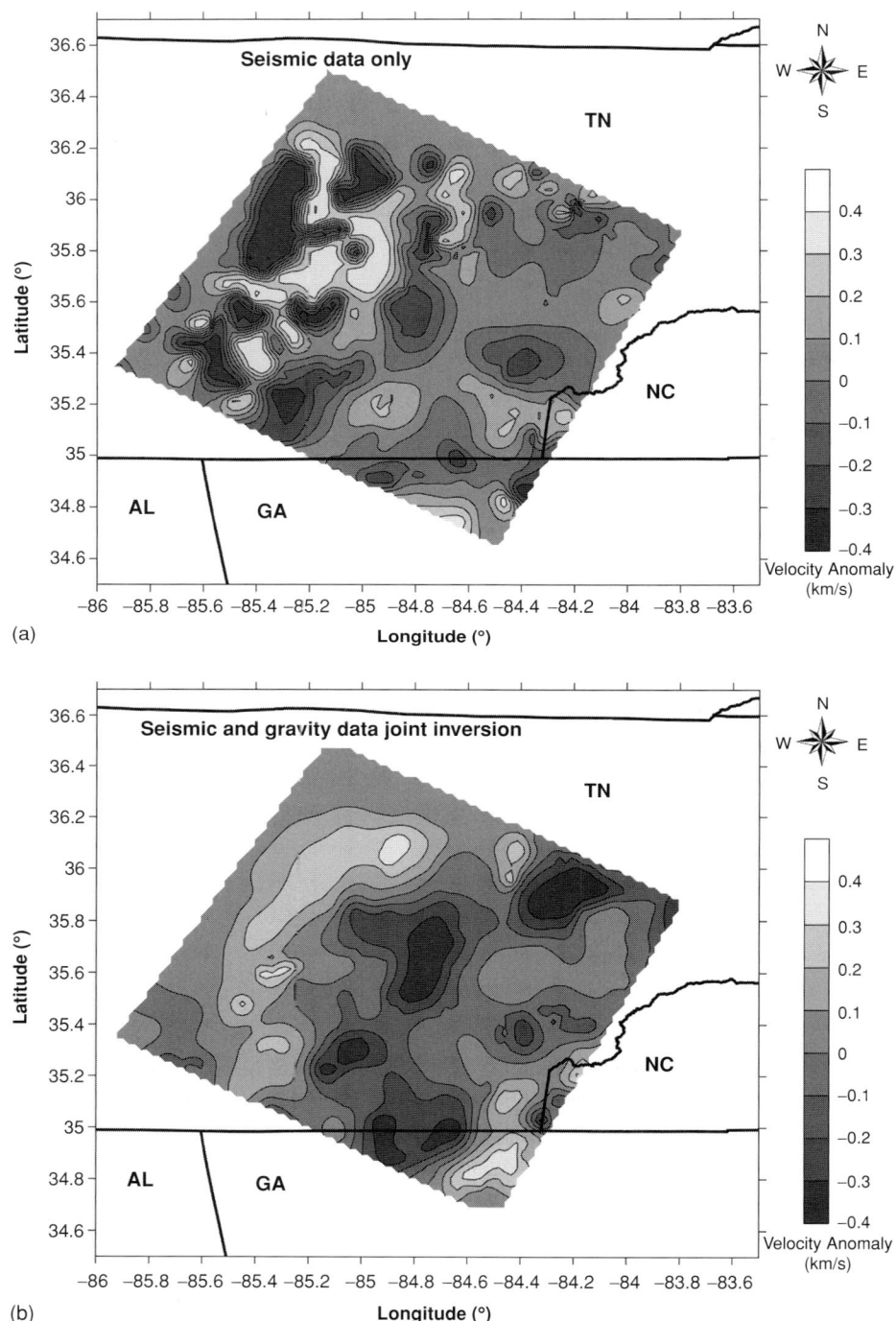

Figure 7.5 Joint inversions of gravity and seismic travel time data to smooth and enhance seismic velocity tomography. These results were contoured from the original data used in Kaufmann and Long (1996) (a) Inversion using only seismic data. (b) Inversion using both seismic and gravity data.

probability density functions (PDFs) that are not Gaussian, as well as consideration of uncertainties in the model and a priori PDFs for the density structure.

The functional relation between the density anomalies and the gravity field is non-linear for most real-world density structures and is expressed as in Eq. (7.1) as

$$0 = F(\Delta g, \Delta \rho). \tag{7.41}$$

In the maximum likelihood method, the solution is found by finding the maximum in the joint probability density function (PDF) for Δg and $\Delta \rho$ given the relation between them defined by F. The probability density function is given as

$$P(\Delta g, \Delta \rho \,|\, F). \tag{7.42}$$

The maximum for this PDF gives the most likely solution for the values of $\Delta \rho$. Computation of the PDF is facilitated by Bayes's theorem

$$P(A \,|\, B) = \frac{P(B \,|\, A)\, P(A)}{P(B)}. \tag{7.43}$$

The application of Bayes's theorem in Eq. (7.43) to Eq. (7.42) gives

$$P(\Delta g, \Delta \rho \,|\, F) = \frac{P(F \,|\, \Delta g, \Delta \rho)\, P(\Delta g, \Delta \rho)}{P(F)} \tag{7.44}$$

and where

$$P(\Delta g, \Delta \rho) = P(\Delta g \,|\, \Delta \rho)\, P(\Delta \rho), \tag{7.45}$$

the relation is

$$P(\Delta g, \Delta \rho \,|\, F) = \frac{P(F \,|\, \Delta g, \Delta \rho)\, P(\Delta g \,|\, \Delta \rho)\, P(\Delta \rho)}{P(F)}. \tag{7.46}$$

The three terms on the right-hand side of Eq. (7.46) represent, respectively, the probability that the model chosen is correct for the gravity and density structure, the probability that the computed gravity values are the observed values given the density structure, and the probability that the density structure is realistic. The $P(F)$, which is not needed to find the maximum value, is a normalization factor based on the probability of a given model.

With gravity data, the model consists of the equations used to compute gravity data given a density structure and the geometry used to set up those equations. Because the equations for computing theoretical gravity are exact, the uncertainty enters through the choice of the shapes of the anomalies and in any computational approximations that are employed to reduce computational effort. An example would be the location of a simple fault at depth that would be assigned a probability distribution for its preferred location, perhaps based on near-surface geology. Also, a probability distribution function could be introduced by using point sources instead of exact shapes, or in two dimensions the uncertainty in using two-dimensional equations instead of 2.5-dimensional equations for modeling data along a line. The advantage of gravity data is that as many density elements as are needed can be introduced so that the probability of the model not being correct could be made negligibly small by increasing the number of elements. The obvious difficulty is that an increased number of elements increases computation time.

The PDF of the observed gravity is primarily determined by the uncertainty in individual gravity observations. For gravity data, the uncertainty in the value would be the combined uncertainties of the precision of the measurement and the geologically induced noise at wavelengths shorter than the station spacing. For closely spaced data, the measurement precision would dominate. For widely spaced data the uncertainty for data inversion should be computed from an estimate of the autocovariance function at a distance comparable to the station spacing (see Chapter 4). The use of the autocovariance function would help to eliminate the influence of small-scale anomalies near one station on the interpretation of larger structures. For the data, the PDF would look very much like a normal distribution, with probabilities decaying rapidly from the observed gravity value. For each gravity station, the probability given a particular model would decrease with an increase in the difference between the observed and computed gravity value. This component in Eq. (7.46) has the same effect as minimizing the weighted least square error. For a sequence of gravity data points the probability can be expanded

$$P(\Delta g \,|\, \Delta \rho) = P(\Delta g_1, \Delta g_2 \cdots |\, \Delta \rho), \tag{7.47}$$

and recognizing the conditional probability among the gravity data, this expands to

$$P(\Delta g \,|\, \Delta \rho) = P(\Delta g_1 \,|(\Delta g_2, \Delta g_2, \Delta g_2 \cdots)|\, \Delta \rho)\, P(\Delta g_2, \Delta g_2, \Delta g_2 \cdots |\, \Delta \rho). \tag{7.48}$$

However, the gravity data PDF for equally spaced data will depend on the precision and the uncertainty introduced by the variations in the gravity field with wavelengths shorter than the data separation. Hence, the conditional probabilities will be approximately the same and dependent on the autocovariance function. Generally, the gravity data probabilities can be computed as independent terms giving the simpler relation

$$P(\Delta g \,|\, \Delta \rho) = P(\Delta g_1 \,|\, \Delta \rho)\, P(\Delta g_2 \,|\, \Delta \rho)\, P(\Delta g_3 \,|\, \Delta \rho) \cdots. \tag{7.49}$$

If there exists any dependence of one gravity value on the others, it is taken up in the use of the autocovariance function to determine the PDF for more widely spaced stations.

The development of the PDF for the density structure can be more interesting. This is where limits on the character and distribution of densities are defined and introduced into the inversion for density structure. The density PDF is the element that introduces most of the interpreter's intuition into a formal inversion. Thus, it is the element that makes the maximum likelihood technique different from the conventional methods based on a least square error. The most obvious condition is that densities are within a specific range for given geologic rocks. In the simplest case densities are restricted to a finite range of values, for example, any density between 2.5 g/cm^3 and 3.0 g/cm^3 could be used for crustal rocks and densities outside this range given very low probability. A more common probability density function illustrated in Figure 7.6 would be one that gives mafic rocks a density of 2.9 $\pm$0.1, granitic rocks a density of 2.67 $\pm$0.05 and sediments at 2.4 $\pm$0.2. All other densities would have a lower probability, as illustrated in Figure 7.6. The relative abundance of these types of rocks would determine the relative heights of the corresponding peaks in the PDF. Hence, the density PDF could be adjusted to fit the distribution of rock density in the study area. The effects of constraining the density values are analogous to the criteria of minimum length, except it is a minimum deviation from what would be expected in a study

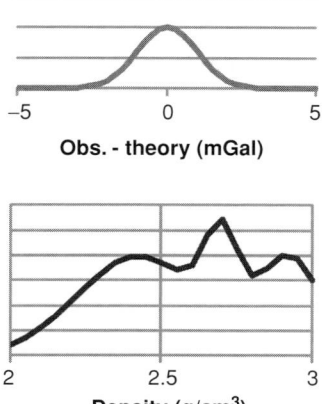

Figure 7.6 PDFs for distribution of densities in crustal rocks that are a mixture of granite composition and basaltic composition intrusive. The PDF for a gravity point typically satisfies a Gaussian distribution.

area. There are also conditional probabilities that can be introduced. As with the probability for the gravity data, the densities can be expanded in the form

$$P(\Delta\rho) = P(\Delta\rho_1, \Delta\rho_2, \Delta\rho_3, \ldots)$$
$$P(\Delta\rho) = P(\Delta\rho_1 \,|\, \Delta\rho_2, \Delta\rho_3, \ldots)\, P(\Delta\rho_2, \Delta\rho_3, \ldots). \tag{7.50}$$

In this form the probability for densities can be increased or decreased depending on whether adjacent values are similar or different, respectively. This constraint has the same influence on the solution for density structure as the flattening matrix in the least squares method, but more flexibility in defining the character of the difference. A PDF that minimizes the differences in model parameters for a given azimuth can be introduced for structures that have a directional component.

In the maximum likelihood method, the joint probability defined as the cost function does not necessarily decay uniformly from the best solution. It can contain secondary maxima. Consequently, the search methods of the minimum square error solution techniques could find secondary maxima that are not the best solution. In general, a minimum square error solution is appropriate only in cases where the probability density functions are all normal Gaussian distributions. In order to find the maximum likelihood solution, individual model components need to be systematically varied in order to map out the joint probability in the multi-parameter model space (one dimension for each model parameter). As an illustrative example, Figure 7.7 shows three gravity values and one density for a model. As the density varies both the theoretical gravity and the likelihood change. The maximum likelihood solution in this example occurs when the density corresponds to a high peak, not the lower value that also is a local maximum in the joint probability. Methods that formalize the search process for the maximum likelihood solution include Monte Carlo and genetic algorithms. These are general search techniques. In the Monte Carlo method, models are chosen at random and only those that exceed a defined likelihood of being correct are kept. The composite of these solutions are kept and examined for the final best solution. In the genetic algorithm,

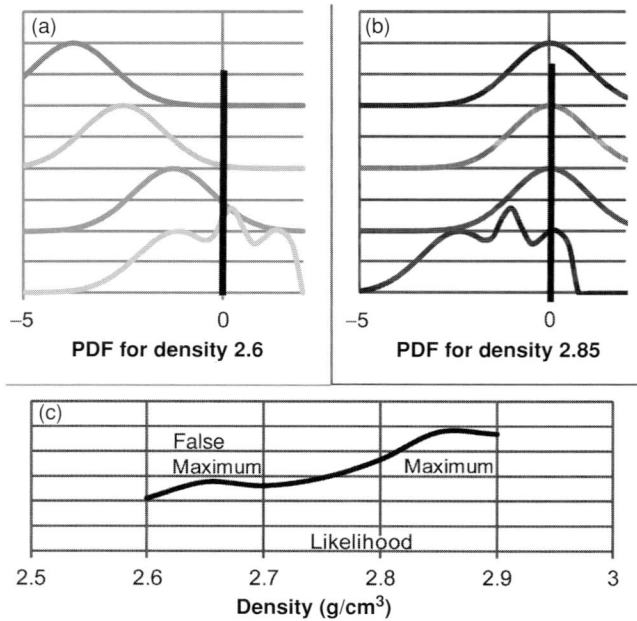

Figure 7.7   (a) PDFs for an unlikely density model. (b) PDFs for a better density model. (c) Likelihood estimates for a sequence of densities showing existence of multiple maxima.

the starting point is a random set of models. The models are ranked in order of the likelihood of being acceptable. Only the best models are kept and used as a basis for new models both by combining and mixing the model parameters and by adding random variations (mutations). Again, only the best are kept and the propagation continued until the likelihood of the model no longer changes. However, once the area of the maximum likelihood has been found, conventional iterative search techniques like Newton's method could be used.

Of all the inversion methods, the maximum likelihood is the most robust, but likely the most difficult to execute because the cost functions and constraints can be difficult to define and because the search techniques can involve many iterations in order to find the best model. The success of the minimum error techniques depend on the accuracy of the model and also require constraints. However, the constraints imposed by the design of the model can lead to instabilities in the analysis and unrealistic solutions for the model. Regardless of the inversion method chosen, most studies begin with the iterative direct modeling approach. With iterative direct modeling, the analyst guesses at a solution and compares it with the data. Then, the guess is modified using the preconceived ideas of the analyst in order to improve the fit with the data. Hence, iterative direct modeling is no more than a highly constrained maximum likelihood approach unlikely to discover the best density model.

8 Experimental isostasy

8.1 The isostatic reduction

The complete Bouguer anomaly has the pronounced effect of removing the attraction of the mass between the point of observation and mean sea level or another convenient datum for reduction. In mountainous regions with an average elevation of 600 m the mass removed introduces Bouguer anomalies on the order of −67 mGal. According to the principle of isostasy, these negative Bouguer anomalies are attributed to the zones of anomalous negative mass below the datum that compensate and support the surface load. The thicker mountainous crust is in effect floating on the denser asthenosphere. In the isostatic reduction the surface mass is not removed as it is in the Bouguer anomaly. Instead, the surface mass is moved to below the datum of reduction. The free air reduction is equivalent to moving the mass into a thin layer at the surface, making the free air reduction a kind of isostatic reduction. In another sense, the isostatic reduction may be viewed as the removal of the topography and the removal of the compensating mass according to a particular model for the compensation. In general, the distribution of compensating mass is unknown. In this chapter we discuss the character of the relation between topography and the Bouguer gravity anomaly and the determination of the distribution of compensating mass.

8.2 The isostatic response function

The isostatic response function is an abstract quantity that cannot be measured directly. Conceptually, it is the Bouguer gravity anomaly resulting from the load of an unrealistic very narrow and tall mountain in isostatic equilibrium. Consequently, the isostatic response function can be a complicated consequence of the elastic strength and composition of the affected rocks. Also, over time, viscoelastic deformation of the underlying rocks, inelastic deformation like faulting, and elastic deformation under stress make the isostatic response function a complex function of time and tectonic history. The Pratt–Hayford and Airy–Heiskanen compensation models (see Chapter 2 and Heiskanen and Moritz, 1967) for computation of the isostatic anomaly illustrate classical models for the expected character of the isostatic response function. In Figure 8.1 a point load corresponding to these models of isostatic compensation is approximated by a 1.0-km high topographic feature with a width of 20 km. In Figure 8.2 the Bouguer anomaly, that is the approximated isostatic

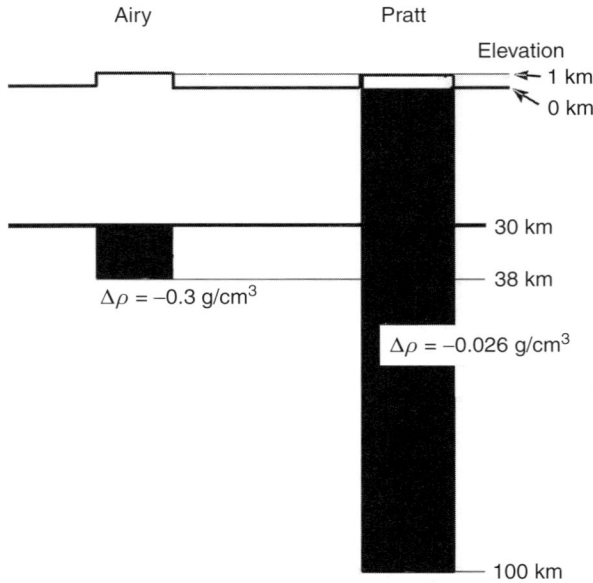

Geometrical interpretation of the Airy model and Pratt model for isostatic compensation.

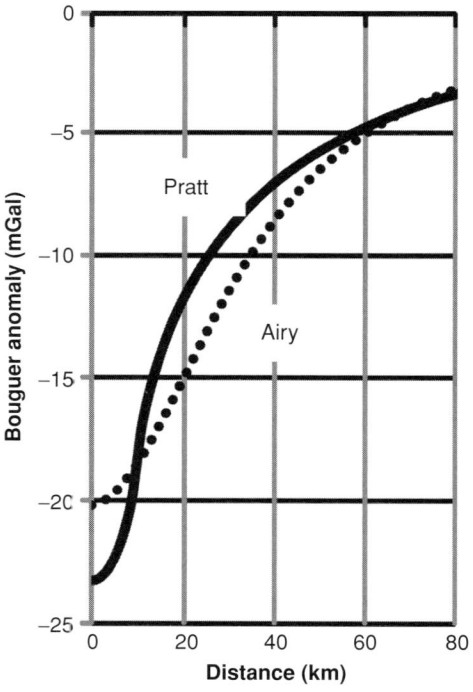

Isostatic response functions for Airy and Pratt models for isostatic compensation.

response function, for these two models are similar, supporting the observation that the isostatic anomaly computed using either of these two models is very similar. In these models, the negative density anomaly of the compensating mass is placed directly below the topographic load. When the lateral distribution of compensating mass is restricted as it is in Figure 8.1, the compensation is called local compensation. Local compensation is physically a reasonable approximation only when the width of the topographic load is greater than the depth to the compensating mass, a condition clearly violated by our point-load mountain. The difference in the isostatic response functions in Figure 8.2 does suggest that if the response function can be defined, one can obtain some indication of the depth of compensation. The Pratt model at short distances is more negative and decreases more rapidly than the Airy model, in response to a portion of the compensating mass being placed at shallow depths. At greater distances, the Pratt model gives smaller anomalies than the Airy model. This is a consequence of placing some compensation at greater depths.

The point-load mountain is an abstract model that would not be expected to exist in nature. Isostatic equilibrium, which assumes a constant pressure at the full depth of compensation, does not guarantee a constant pressure at intervening depths. Two adjacent columns with different vertical distributions of density can generate anomalous and non-vertical stresses at intervening depths. These stresses add to any residual stresses from past tectonic activity, and stresses from contemporary tectonic activity. The resulting lateral variations in stress will either be sustained by the elastic strength of the rock or in weak rocks cause deformation, leading to a reduction in the stress. The traditional application of analyses of the isostatic response is in the study of the dynamics of the crust, and in particular, the oceanic crust because of its more predictable geologic structure. Dramatic examples of the principal of isostasy on continents and the response of the crust to surface loads can be found in analyses of depositional basins and off-shore deltas. The isostatic response function should include the lateral redistribution of mass that will reduce lateral stress differences to levels that can be sustained by the elastic strength of the rocks. Also, the isostatic response function should contain the diffusion the topographic load by elastic stress. When the Pratt model is used to compute isostatic anomalies, a depth of compensation on the order of 100 km is generally assumed. In reality, we know that compensation is largely complete at depths near the Moho, or around 30 km. In general, then, the isostatic response function with time after placement of the load on the surface will be related to both a lateral and depth variation in compensation. Watts (2001) presents a more complete analysis of the components contributing to isostatic compensation and its relation to flexure of the lithosphere. In this chapter we present the fundamental problems related to isostatic compensation and its application to gravity data analysis and interpretation.

8.3 Determination of the isostatic response function

The isostatic response function, if it exists as a unique property of the crust, in theory can be computed from the Bouguer gravity anomaly and topography only after making some significant assumptions. The simplest model would be an isotropic crust with topographic loads only at the surface. These assumptions as presented in Dorman and Lewis (1970) are as follows:

(1) The isostatic response function is isotropic; that is, the response for a point load is independent of azimuth. This means that the isostatic response function $R(r)$ is a function of only the distance, r. Also, we assume that the compensating mass $\Delta\rho\,(\zeta, r)$ for a point load is independent of azimuth. The compensating density for a point load may also be a function of depth. This assumption is realistic for averages over large areas and crustal features like volcanic islands on an oceanic crust. However, in linear orogenic belts on continental crust the isostatic response parallel to the structure has been found to differ from the isostatic response perpendicular to the strike of the structures.

(2) The compensating mass is a function of position (ξ, η) and depth ζ below the topography. We define the density structure with the function $\Delta\rho'\,(\varsigma, \eta, \zeta)$.

(3) The compensating mass $\Delta o'\,(\varsigma, \eta, \zeta)$ is linearly related to the weighted average of the topographic load. The weighted average is expressed in integral form as

$$\Delta\rho'\,(\xi, \eta, \zeta) = \iint \Delta\rho\,(\zeta, r)\,h(x, y)dxdy, \tag{8.1}$$

where

$$r = \left((x - \xi)^2 + (y - \eta)^2\right)^{1/2}. \tag{8.2}$$

Equation (8.1) can be recognized as the convolution of topography with the compensation for a point load. Convolution may be written in short form as $\Delta\rho' = \Delta\rho * h$. The topography, $h\,(x, y)$, is known, but the compensating mass and the density distribution for a point topographic load are both unknown.

(4) The assumption of linearity above assumes that the isostatic response function for a point load is linearly related to the height of the topography. This assumption requires that the crustal response, that is the natural placement of mass to achieve isostatic equilibrium at all depths, is independent of the stresses generated by the load. This assumption will have difficulty in a crustal rheology that is non-linear. A response that varies with the magnitude of the stress will make the isostatic response function differ for different mountain elevations.

(5) The isostatic response function is independent of time. During the active periods of continental rifting or collision, changes in temperature, heat flow and fluid content can change the strength of the crust and generate a response significantly different from that found for a cooler, more rigid crust. The variations in the viscosity of the mantle and crust will also contribute to variations in the isostatic response function with time.

For the classical isostatic reductions, the function $\Delta\rho$ is assumed in order to compute the isostatic anomalies. The actual distribution of the compensating density is generally unknown and cannot be uniquely determined from the observed alone. However, we can express the Bouguer gravity anomalies in terms of the unknown density distribution and use this relation to find the isostatic response function. The vertical component of gravitational attraction, the Bouguer anomaly, can be expressed as

$$\Delta g_z\,(x, y, z = 0) = G \iiint \frac{\zeta \Delta \rho'\,(\xi, \eta, \zeta)}{\left((x - \xi)^2 + (y - \eta)^2 + \zeta^2\right)^{3/2}} d\xi\, d\eta\, d\zeta. \tag{8.3}$$

From Eq. (8.1) above we can substitute the convolution of elevation with the compensation for a point load for the compensation $\Delta \rho'$. Using the appropriate dummy variables, (x', y') Eq. (8.3) can be written as

$$\Delta g_z(x, y, z = 0) = G \iiiint \frac{\zeta \Delta \rho'(\xi - x', \eta - y', \zeta)h(x', y')}{\left((x - \xi)^2 + (y - \eta)^2 + \zeta^2\right)^{3/2}} dx'\, dy'\, d\xi\, d\eta\, d\zeta. \tag{8.4}$$

Then, by reversing the order in integration and substituting ξ' for $\xi - x'$ and η' for $\eta - y'$ Eq. (8.3) can be written as

$$\Delta g_z(x, y) = \iint \left[G \iiint \frac{\Delta \rho(\xi', \eta', \zeta)\zeta\, d\xi'\, d\eta'\, d\zeta}{\left((x - x' - \xi')^2 + (y - y' - \eta')^2 + \zeta^2\right)^{3/2}} \right] h\,(x', y')\, dx'\, dy', \tag{8.5}$$

where the portion in brackets is the isostatic response function, $R\,(x - x', y - y')$. Hence, Eq. (8.3) for the Bouguer gravity can be written as a convolution of the topography with the isostatic response function for a point load

$$\Delta g_z(x, y) = \iint R(x - x', y - y')h(x', y')dx'\,dy'. \tag{8.6}$$

The convolution may be written in short form as

$$\Delta g_z = R^* h. \tag{8.7}$$

The gravity anomaly in Eq. (8.6) represents the Bouguer anomaly associated with isostatic compensation and the topography in Eq. (8.5) represents only the topographic load. The gravity anomalies from shallow and small geological structures may be viewed as noise. They are unrelated to the isostatic response and more correctly represent contributions to the load not represented in the topography. The effects of short-wavelength density anomalies at shallow depths could be removed by filtering and equivalent topographic loads added to the existing topography to compute the isostatic response function. Alternatively, the assumed azimuthally symmetry of the response function could be used to average out the effects of geologic structures. The difference between these two approaches is in the nature of the way noise is removed. If the gravity anomaly from the isostatic response is set equal to the sum of the observed Bouguer gravity Δg_B and anomalies from geologic structures Δg_g Eq. (8.7) can be written as

$$\Delta g_z(x, y) = \Delta g_B - \Delta g_g = R^* h. \tag{8.8}$$

The planar approximation, which assumes that the curvature of the Earth's surface is negligible over the area of interest, is sufficient for most studies of regional isostatic response. In the planar approximation, Eq. (8.8) can be solved by use of the two-dimensional Fourier transform. In the wavenumber domain of the Fourier transform, the isostatic response function is usually referred to as the admittance function that relates gravity to topography. In the transform domain, the convolution becomes a multiplication and Eq. (8.8) can be written as

$$(\Delta g_B(k_x, k_y) - \Delta g_g(k_x, k_y)) = R(k_x, k_y)h(k_x, k_y). \tag{8.9}$$

Provided no component of the transform of $h(x,y)$ is zero, the transform of the isostatic response function (the admittance) can be expressed as

$$R(k_x, k_y) = \frac{(\Delta g_B(k_x, k_y) - \Delta g_g(k_x, k_y))}{h(k_x, k_y)}, \tag{8.10}$$

where the components of Eq. (8.8) have been transformed to the wavenumber domain (k_x, k_y). Because of the assumption of an isotropic crust, $R(x, y)$ is real and symmetric about the origin, the transform $R(k_x, k_y)$ is also real and symmetric about its origin ($k_x = 0, k_y = 0$). The variations associated with geological anomalies would thus be expected to average to zero if Eq. (8.10) is averaged over all values of constant wavenumber $|k| = (k_x^2 + k_y^2)^{1/2}$. On averaging over all values of constant $|k|$, or more commonly over a finite range of $|k|$, Eq. (8.10) reduces to the real part of the average value for the ratio of the Bouguer anomalies to the elevation in the form

$$R(|k|) = R_e \left(\frac{\langle B(|k|) - G(|k|) \rangle}{\langle H(|k|) \rangle} \right) \sim R_e \left(\frac{\langle B(|k|) \rangle}{\langle H(|k|) \rangle} \right), \tag{8.11}$$

where the angle brackets indicate averages and R_e ($-$) takes the real part. In effect, Eq. (8.11) uses azimuthally symmetry to reduce the dimension from two to one. The admittance function at this point can then be obtained by replacing each value in $R(k_x, k_y)$ by the averaged value for each $|k|$ from Eq. (8.11) and performing the two-dimensional inverse Fourier transform.

As an alternative to the inverse Fourier transform, the Fourier transform Eq. (8.5) could be converted to polar coordinates and the average computed by integrating first over azimuth. The result is an integral expression for the isostatic response function as a function of distance from the point load

$$r(\gamma) = \int |k| R(|k|) J_0(2\pi |k|) d|k|, \tag{8.12}$$

where J_0 is the zeroth-order Bessel function.

Smoothing as in Eq. (8.11) is an essential part of the analysis of real data. In two dimensions, the average is over a given wavelength. For one-dimensional lines, averaging can only partially be achieved by using multiple lines perpendicular or parallel to the dominant structure. This does not always average out anomalous geologic structures. In continental applications in particular, erosion and sedimentation can introduce significant phase differences between the Bouguer and elevation spectra. For example, a mountain range could be significantly eroded on one side leaving compensation at depth and an asymmetrical apparent response function. Also, linear tectonic belts representing differing crustal composition can introduce noise in the Bouguer anomalies that is too large to remove by averaging.

The ability to extract an isostatic response function depends strongly on how smoothing is applied. In Eq. (8.11) soothing was over the azimuth, an option not available for one-dimensional data. Because white noise is a constant in the transform domain, the level of noise can be recognized in the spectra and removed by subtracting a constant defined by the noise level. Also, because the length of a line of data should be significantly longer than the response function, smoothing over a range of wavelengths will reduce the noise and effectively return only the significant portion of the isostatic response function. The admittance function may also be computed from Eq. (8.9) if instead it is post-multiplied by the complex conjugate of the elevation. The relation for lines of data in one-dimensional data becomes

$$(\Delta g_B(k) - \Delta g_g(k))h^*(k) = R(k)h(k)h^*(k), \tag{8.13}$$

which on solving for the admittance function becomes

$$R(k) = \frac{(\Delta g_B(k) - \Delta g_g(k))h^*(k)}{h(k)h^*(k)}. \tag{8.14}$$

The advantage of using the covariance is that wavenumbers in the Bouguer anomalies that are out of phase with the elevation are suppressed. A disadvantage is that evidence for asymmetry is lost.

The inverse transform of the admittance should give the isostatic response function. However, because both the Bouguer anomalies and the elevation data are referenced to a somewhat arbitrary datum, the longer-wavelength levels typically have to be adjusted. For this reason, and because there are problems with controlling noise at shorter wavelengths, most analysis are performed on the admittance function.

8.4 Interpretation of the isostatic response function

The interpretation of the isostatic response function is the familiar non-unique problem of finding an acceptable density structure for a given gravity anomaly. The interest and utility of the isostatic response function is its direct connection to the tectonic forces of mountain building and basin subsidence. The simplest assumption for the compensating mass is that it is purely local compensation. The isostatic response function is then interpreted as a function of depth, which is equivalent to writing the compensation as a product of a delta function in position and a function of depth

$$\Delta \rho(\zeta, \gamma) = \Delta \rho(\zeta)\delta(\gamma). \tag{8.15}$$

The attraction of an incremental mass in a vertical column can be approximated by the expression for a point mass as

$$R(\gamma) = \frac{G\zeta}{(\gamma^2 + \zeta^2)^{3/2}}. \tag{8.16}$$

Integrating this expression over all depths gives the expression

$$\Delta g_{r_e}(\gamma) = G \int_0^{r_e} \frac{\zeta \Delta\rho(\zeta)}{(\gamma^2 + \zeta^2)^{3/2}} d\zeta. \tag{8.17}$$

The solution for $\Delta\rho(\zeta)$, the density as a function of depth, can be found numerically from this integral expression for the isostatic response function or analytically following the technique of Dorman and Lewis (1970). Because this inversion precludes inclusion of a lateral component in the response function, its usefulness is questionable.

8.5 Example computation of the admittance and isostatic response functions

It is instructive to examine terrestrial applications of the solution for the admittance and isostatic response functions in order to illustrate some of the problems and complications in this type of analysis. As a first example a line perpendicular to the regional structure of the southern Appalachian Mountains is examined. Gravity data and elevations along a 50-km wide line extending northwest are smoothed over a 10-km length along the line. These data are shown in Figure 8.3, where the elevation is scaled down to bring out correlations with the gravity anomalies. The data demonstrate a marginal correlation between elevation and Bouguer anomalies that results from tectonic history and differing geologic terrains of this region. The free air anomalies average near zero, but the average has to be taken over distances exceeding 200 km to demonstrate equilibrium. The maximum in the elevation at 425 km is northwest of the maximum in the Bouguer anomaly at 400 km. In this area the topography has been removed through erosion leaving compensated mass at depth. This negative anomaly is associated with an abrupt 10 km increase in crustal thickness Cook et al. (1981), Hawman (2008), and Hawman et al. (2011) give more detail and alternative interpretations. Near 275 km the large positive gravity anomalies are associated with mafic intrusive and extrusive rocks in the Eastern Piedmont. Also, near 100 km there exists the remnants of Mesozoic volcanic activity that also gives a positive gravity anomaly that is independent of the surface topography. In summary, this line is dominated by major geologic structures of the crust and as the free air anomalies suggest the isostatic response function, if it exists, should be on the order of 200 km in width.

For analysis, the data in Figure 8.3 were sampled at increments of 3 km for a total distance of 384 km. In order to eliminate truncation effects in the transform, symmetry of the profile was assumed and the trace length was doubled by reflecting it about a point at 425 km. The log spectra versus log wavenumber are plotted in Figure 8.4a. In Figure 8.4a,

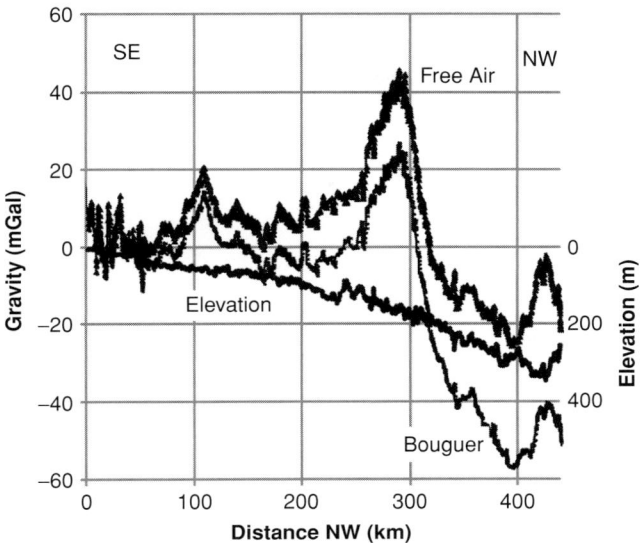

Figure 8.3 Topography and gravity perpendicular to the trend of the southern Appalachian Mountains. The elevation data are divided by −10 to allow easier comparison with the gravity data.

the wavelength has the value of 384 divided by the wavenumber. The admittance function in Figure 8.4a was computed using Eq. (8.11). The geologic components for these data do not average out as suggested in Eq. (8.11). Instead, the geologic term was replaced by white noise that is equivalent to the constant level of the spectra at shorter wavelengths. In order to compute the isostatic response function, the inverse transform of the admittance function, additional corrections were applied. The first four wavenumbers, representing the long-wavelength components are somewhat arbitrary, because the reference for zero in both the elevation and the gravity anomaly are arbitrary. The heavy line in Figure 8.4a shows the adjustment for these long wavelengths. The second correction was to truncate the spectra at wavenumber 32 in order to eliminate the influence of the uncorrelated short wavelengths. These short wavelengths in the admittance function will dominate the isostatic response function if not smoothed out or truncated.

Figure 8.4b shows the resulting isostatic response function. This isostatic response function is misleading because the rapid fall-off with distance suggests that the isostatic response is at depths on the order of 20 km, placing the compensation at mid-crustal depths. This interpretation differs from the implications of the free air anomalies as well as the observation of Lewis and Dorman (1970) for the continental United States. The Lewis and Dorman (1970) result is shown for comparison in Figure 8.4b. The difference is principally in the rate at which the response curves approach or pass through zero. A better interpretation of the southern Appalachian transect is that the geological component of the Bouguer anomalies dominate the response. In this case the isostatic response function does not explain the relation between gravity and topography and explanations should in general be coupled with information on the tectonic history and geologic structures of the region. A similar

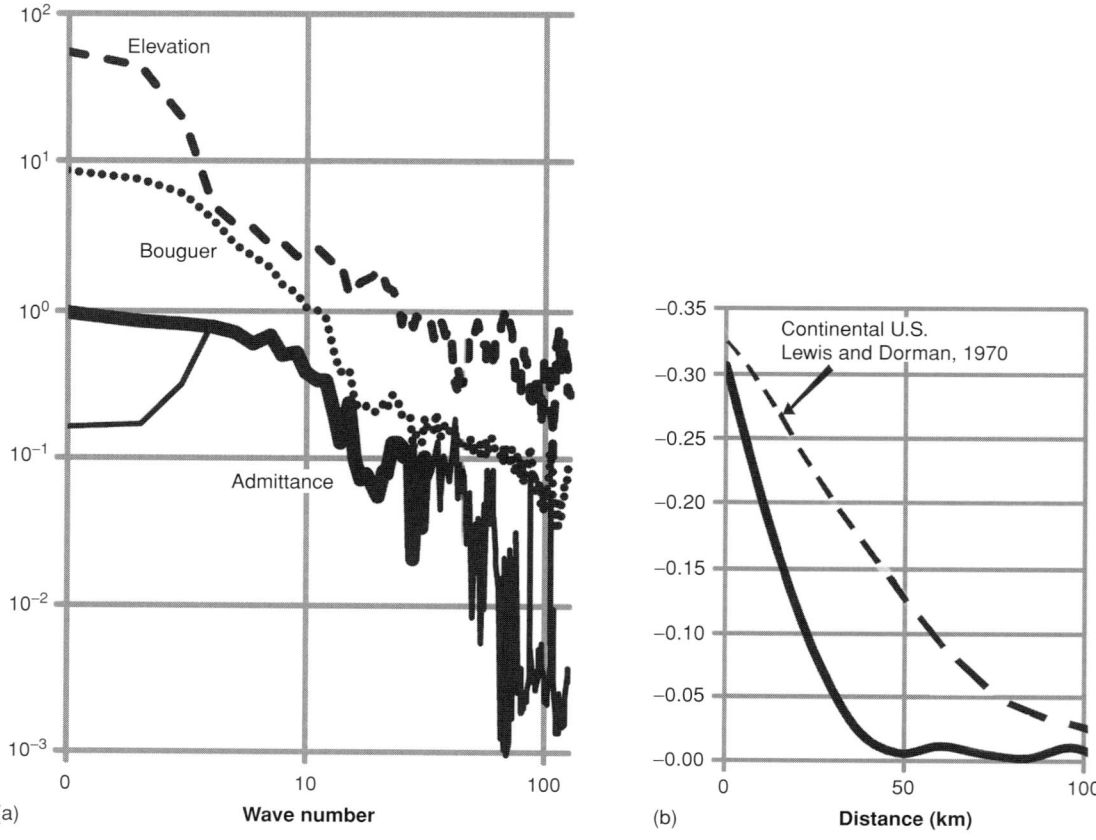

Figure 8.4 (a) Admittance function for the southern Appalachian Mountains and the Fourier transforms of the topography and Bouguer anomalies used in its computation. The heavy line in the admittance shows the corrections applied for long wavelengths. (b) Isostatic response functions computed for the southern Appalachian Mountains from Figure 8.4a and for the continental United States from Dorman and Lewis (1972).

computation for elevation and gravity data taken parallel to the structures would indicate a response more like the Lewis and Dorman (1970) continental Unites States response. Similar results were obtained by Stephenson and Lambeck (1985) for studies on the Australian Continent and by Long and Liow (1986) for different data in the southern Appalachian Mountains.

8.6 Isostasy in coastal plain sediments

Although the isostatic response function works at large scales on continental crust, its application is hindered by the influence of dominating tectonic and geologic influences on the relation between elevation and gravity. However, one could speculate whether at a smaller

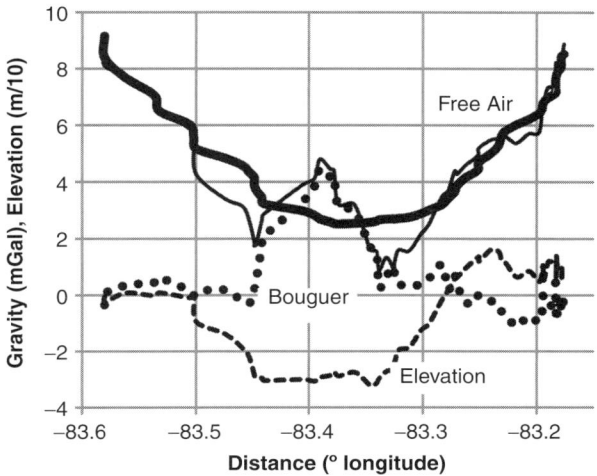

Figure 8.5 Gravity data and topography from the Coastal Plain of Georgia. The heavy line is the smoothed free air used to remove regional gravity anomalies. The line shows the anomalous correlation of elevation with Bouguer anomalies. A Bouguer reduction density of 2.0 g/cm^3 was used in this analysis. The dotted line is the Bouguer anomaly after removal of the regional trend. The dashed line is the elevation.

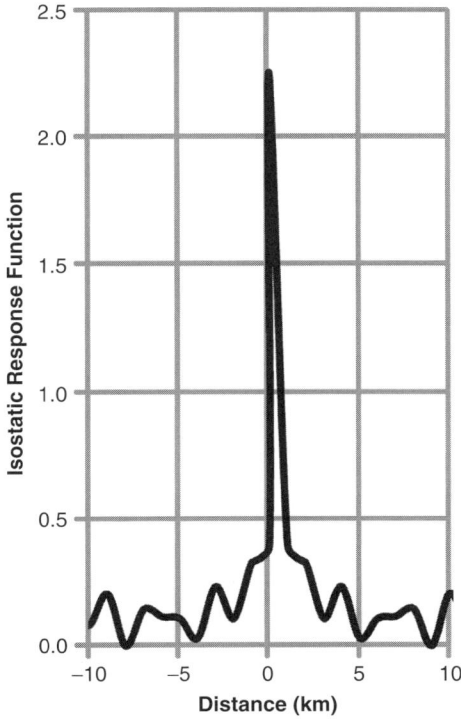

Figure 8.6 Isostatic response function of the line in Figure 8.5. The response is essentially an impulse function.

scale isostasy could be a viable component in the analysis of near-surface features. In particular, in sediment deposition areas where the near-surface rocks are still weak one could speculate on whether the achievement of isostatic equilibrium is an operating mechanism. Figure 8.5 shows a profile of gravity and elevation in coastal plain sediments at a location where there is an eroded river channel. The free air anomalies appear to include anomalies that average out over a distance of 0.1 degrees of longitude (about 10 km) and the averaged values can be used as a proxy for the regional or crustal anomalies. The elevation contrast in the profile is a 30-m deep river channel that correlates with a positive Bouguer anomaly. The model here would appear to be an inverse mountain compensated by excess mass underneath the topographic low. Isostatic response function analysis might allow determination of whether this anomaly is related to isostatic response in coastal plain sediments, or from some other source. The smoothed free air anomalies were used to remove the regional anomalies from the Bouguer anomalies. Both the elevation and Bouguer anomalies were adjusted to have zero values at each end of the profile. For the computation of the admittance function, Eq. (8.14) was used as a means of suppressing incoherence between the elevation and gravity anomalies. The resulting admittance closely approximates white noise and inversion back to the isostatic response function gives an impulse function shown in Figure 8.6. Hence, for this data the elevation and Bouguer anomaly are directly correlated and the compensation, if it exists, would have to be at very shallow depths. A more likely explanation is that the correlation between the topography and Bouguer anomaly in this line is coincidental.

Appendix A Common definitions and equations in potential theory

The analysis of gravity is highly dependent on potential theory. In this Appendix we list common mathematical terms and identities used in this book. A more detailed treatment of potential theory can be found in the classical books by MacMillan (1958) and Blakely (1995)

Fields: A field is that space in which a physical quantity, for example, gravity, temperature, or velocity, is measurable. Fields considered in this text are continuous and have one and only one value associated with each point within the space.

Scalar Field: A scalar field is a single-valued field that is a function only of magnitude at each point, for example, temperature or hydrostatic pressure.

Vector Field: A vector field is a set of "n" functions in "n"-dimensional space that describe magnitude and direction at each point. In three dimensions, each point is described by three functions, one each for the magnitude in the three orthogonal directions. Gravitational attraction is one example.

Tensor Field: A tensor field, or an nth order tensor in 3-dimensional space, has each point described by 3^n functions. In "m" dimensional space each point is described by m^n functions.

Vectors in Cartesian Coordinates: A vector **A** can be expressed as an array of its components in the orthogonal directions of the coordinate system. In three-dimensional space, the conventional nomenclature for the axes are x, y, and z, and the scalar magnitudes of the components of the vector are (A_x, A_y, A_z). The vectors (**i**, **j**, **k**) are the unit vectors in the (x, y, z) directions, respectively. For convenience in n-dimensional space where n may be greater than three the axes are indicated as 1, 2, 3, 4, $-n$) and tensor notation may be used. The scalar magnitude of a vector is written A, or $|\mathbf{A}|$. A vector expressed in terms of the three orthogonal unit vectors is

$$\mathbf{A} = A_x\mathbf{i} + A_y\mathbf{j} + A_z\mathbf{k}. \tag{A.1}$$

The scalar magnitude or length of a vector is also expressed as

$$A = \sqrt{A_x^2 + A_y^2 + A_z^2}. \tag{A.2}$$

Multiplication of a vector by a scalar leaves the direction unchanged and only changes the magnitude. One adds two vectors multiplied by the scalars α and β by using

$$\alpha\mathbf{A} + \beta\mathbf{B} = (\alpha A_x + \beta B_x)\mathbf{i} + (\alpha A_y + \beta B_y)\mathbf{j} + (\alpha A_z + \beta B_z)\mathbf{k}. \tag{A.3}$$

Multiplication by a scalar is distributive, so that

$$\alpha(\mathbf{A} + \mathbf{B}) = \alpha\mathbf{A} + \alpha\mathbf{B} \tag{A.4}$$

and vector addition is commutative

$$\mathbf{A} + \mathbf{B} = \mathbf{B} + \mathbf{A}. \tag{A.5}$$

Inner product: The inner product or dot product reduces the order of a tensor by one. The inner product of two vectors yields a scalar. The inner product of two vectors is given by

$$\mathbf{A} \cdot \mathbf{B} = A_x B_x + A_y B_y + A_z B_z = \alpha. \tag{A.6}$$

If $\mathbf{A}$ and $\mathbf{B}$ are vector fields, the magnitude of the scalar field α is given by the product of the length of each vector and the cosine of the angle θ between the vectors

$$\alpha = AB \cos \theta. \tag{A.7}$$

The inner product of a vector with itself gives the square of its length. The inner product of orthogonal vectors, vectors at right angles to each other, is zero by definition because the cosine of 90 degrees, the angle between them, is zero.

Cross product: The cross product or vector product of two vectors yields a vector that is orthogonal to both of the original vectors. The cross product of $\mathbf{A}$ with $\mathbf{B}$ defines a vector $\mathbf{C}$ that is perpendicular to both $\mathbf{A}$ and $\mathbf{B}$

$$\mathbf{C} = \mathbf{A} \times \mathbf{B} = \begin{vmatrix} \mathbf{i} & \mathbf{j} & \mathbf{k} \\ A_x & A_y & A_z \\ B_x & B_y & B_z \end{vmatrix}. \tag{A.8}$$

Note that $(\mathbf{A} \times \mathbf{B}) = -(\mathbf{B} \times \mathbf{A})$. The magnitude of $\mathbf{C}$ is given by the length of each vector and the sine of the angle between the vectors.

$$|\mathbf{C}| = |\mathbf{A}||\mathbf{B}| \sin AB. \tag{A.9}$$

Tensor product: The tensor product or outer product increases the order of the tensor. The outer product of two vectors (first-order tensors) gives a second-order tensor, $\mathbf{T}$.

$$\mathbf{T} = \mathbf{A} \otimes \mathbf{B} = \begin{vmatrix} \mathbf{i} & \mathbf{j} & \mathbf{k} \end{vmatrix} \begin{vmatrix} A_x B_x & A_x B_y & A_x B_z \\ A_y B_x & A_y B_y & A_y B_z \\ A_z B_x & A_z B_y & A_z B_z \end{vmatrix} \begin{vmatrix} \mathbf{i} \\ \mathbf{j} \\ \mathbf{k} \end{vmatrix}. \tag{A.10}$$

The ∇ operator: The ∇ operator is given by

$$\nabla = \mathbf{i} \frac{\partial}{\partial x} + \mathbf{j} \frac{\partial}{\partial y} + \mathbf{k} \frac{\partial}{\partial z}. \tag{A.11}$$

Gradient: A gradient is defined when ∇ is applied to a scalar field ϕ. The gradient is a vector indicating the direction and steepness of the decrease in values of a scalar.

$$\nabla \phi = \mathbf{i} \frac{\partial \phi}{\partial x} + \mathbf{j} \frac{\partial \phi}{\partial y} + \mathbf{k} \frac{\partial \phi}{\partial z}. \tag{A.12}$$

Given a scalar function $\phi(x, y, z)$, the equation $\phi(x, y, z) = \phi_0$ where ϕ_0 is a constant defines a surface in a continuously varying scalar field. If $\phi(x, y, z)$ is a potential, ϕ_0 is an equipotential surface. The gradient, $\nabla \phi$, on the surface is perpendicular to the surface defined by the constant ϕ_0.

Divergence: The divergence, a scalar function, is obtained when the ∇ operator is applied to a vector using the inner product.

$$\nabla \cdot \mathbf{F} = \frac{\partial F_x}{\partial x} + \frac{\partial F_y}{\partial y} + \frac{\partial F_z}{\partial z}. \tag{A.13}$$

When the divergence is applied to a second-order tensor, the result is a vector (a first-order tensor). The divergence is an inner product and reduces the order by one. $\nabla \cdot \mathbf{F} = \mathbf{0}$ implies that $\mathbf{F}$ was derived from a vector field $\mathbf{A}$ such that $\mathbf{F} = \nabla \times \mathbf{A}$.

Curl: The Curl is defined by ∇ applied to a vector using the cross product

$$\nabla \times \mathbf{F} = \mathbf{Curl}(\mathbf{F}) = \begin{vmatrix} \mathbf{i} & \mathbf{j} & \mathbf{k} \\ \frac{\partial}{\partial x} & \frac{\partial}{\partial y} & \frac{\partial}{\partial z} \\ F_x & F_y & F_z \end{vmatrix}. \tag{A.14}$$

$\nabla \times F = 0$ implies that F was derived from a scalar function U, such that $F = \nabla U$. The outer product of ∇ with a vector gives a second-order tensor with components

$$\nabla \otimes B = \begin{vmatrix} \mathbf{i} & \mathbf{j} & \mathbf{k} \end{vmatrix} \begin{vmatrix} \frac{\partial}{\partial x} B_x & \frac{\partial}{\partial x} B_y & \frac{\partial}{\partial x} B_z \\ \frac{\partial}{\partial y} B_x & \frac{\partial}{\partial y} B_y & \frac{\partial}{\partial y} B_z \\ \frac{\partial}{\partial z} B_x & \frac{\partial}{\partial z} B_y & \frac{\partial}{\partial z} B_z \end{vmatrix} \begin{vmatrix} \mathbf{i} \\ \mathbf{j} \\ \mathbf{k} \end{vmatrix}. \tag{A.15}$$

Conservative Field: A scalar field is conservative when the line integral from point M to point N is independent of the path. If N and M are the same point, then an equivalent statement is that if the line integral goes to zero around any closed path then the field is conservative.

$$\phi(M) = \phi(N) + \int_M^N \nabla\phi \cdot d\mathbf{r}. \tag{A.16}$$

Laplacian: The Laplacian is defined by

$$\nabla \cdot \nabla = \nabla^2 = \frac{\partial^2}{\partial x^2} + \frac{\partial^2}{\partial y^2} + \frac{\partial^2}{\partial z^2}. \tag{A.17}$$

Vector Identities: U and V are scalar functions. **F**, **A**, and **B** are vectors.

$$\nabla \times (\nabla V) = 0 \tag{A.18}$$

$$\nabla \cdot (\nabla \times \mathbf{F}) = 0 \tag{A.19}$$

$$\nabla^2 \phi = \nabla \cdot (\nabla \phi) \tag{A.20}$$

$$\nabla(UV) = U\nabla V + V\nabla U \tag{A.21}$$

$$\nabla \times \nabla \times \mathbf{A} = \nabla(\nabla \cdot \mathbf{A}) - \nabla^2 \mathbf{A} \tag{A.22}$$

$$\nabla \cdot (\mathbf{A} \times \mathbf{B}) = \mathbf{B} \cdot (\nabla \times \mathbf{A}) - \mathbf{A} \cdot (\nabla \times \mathbf{B}) \tag{A.23}$$

$$\nabla \cdot (U\mathbf{A}) = U\nabla \cdot \mathbf{A} + (\mathbf{A} \cdot \nabla)U \tag{A.24}$$

$$\nabla \times (U\mathbf{A}) = U(\nabla \times \mathbf{A}) + (\nabla U) \times \mathbf{A} \qquad (A.25)$$

$$\nabla \times (U\mathbf{A}) = U(\nabla \times \mathbf{A}) + (\nabla U) \times \mathbf{A} \qquad (A.26)$$

$$\nabla \times (\mathbf{A} \times \mathbf{B}) = \mathbf{A}(\nabla \cdot \mathbf{B}) - \mathbf{B}(\nabla \cdot \mathbf{A}) + (\mathbf{B} \cdot \nabla)\mathbf{A} - (\mathbf{A} \cdot \nabla)\mathbf{B} \qquad (A.27)$$

$$\nabla \times (\mathbf{A} \times \mathbf{B}) = \mathbf{A}(\nabla \cdot \mathbf{B}) - \mathbf{B}(\nabla \cdot \mathbf{A}) + (\mathbf{B} \cdot \nabla)\mathbf{A} - (\mathbf{A} \cdot \nabla)\mathbf{B}. \qquad (A.28)$$

Integral identities:

$$\iint (\nabla \times \mathbf{F}) \cdot d\mathbf{s} = \int \mathbf{F} \cdot d\mathbf{l} \qquad (A.29)$$

$$\nabla \times \mathbf{F} = \lim_{ds \to 0} \frac{\mathbf{F} \cdot d\mathbf{l}}{d\mathbf{s}} \qquad (A.30)$$

$$\int_{v} \nabla \cdot \mathbf{F} dv = \int_{s} \mathbf{F} \cdot d\mathbf{s} = \int_{s} F_n ds \qquad (A.31)$$

$$\nabla \cdot \mathbf{F} = \lim_{v \to 0} \frac{1}{v} \int_{s} \mathbf{F} \cdot d\mathbf{s} \qquad (A.32)$$

$$\iint_{s} U d\mathbf{s} = \iiint_{v} \nabla U dv \qquad (A.33)$$

$$\iiint_{v} (\nabla \times \mathbf{A}) dv = \iint_{s} (d\mathbf{s} \times \mathbf{A}) \qquad (A.34)$$

$$\int_{l} (\mathbf{A} \cdot \mathbf{B}) dl = \iint_{s} (\nabla \mathbf{A} \cdot \mathbf{B}) \times d\mathbf{s}. \qquad (A.35)$$

Appendix B **Glossary of symbols**

α	measure of an angle
α	weight factor in a weighted average
α_{ij}	the weight of the ith component for the jth value
β	latitude
γ	theoretical gravity from an ellipsoid fitted to the geoid
γ_a	theoretical gravity at the equator
γ_b	theoretical gravity at the pole
Δ	a difference operator on the symbol that follows
δ	indicator of incremental change in symbol that follows
ε_i	indicates the error of the ith measure
ζ	the z coordinate of attracting matter
η	the y coordinate of attracting matter
θ	spherical coordinate corresponding to longitudinal coordinate
κ	mass per unit area of a thin sheet
λ	longitude coordinate, wavelength, Lagrange multipliers
μ_i	mean value of the ith component.
ξ	the x coordinate of attracting matter
(ξ, η, ζ)	coordinates of attracting matter corresponding to x, y, z, respectively
ρ	density
$\Delta\rho$	density anomaly
σ	surface area, $d\sigma$ in incremental surface area.
τ	shift distance in covariance and correlation coefficient integrals
ϕ	spherical coordinate corresponding to the latitudinal coordinate, on an ellipsoidal Earth
ω	rate of rotation of the Earth, frequency in Fourier transform frequency domain
a	acceleration of moving body or anomalous mass with components (a_x, a_y, a_z)
a	equatorial radius of Earth, general distance measure
a_i, b_i	general constants in a function expansion, for example Fourier series
A	cross-sectional area
b	polar radius of Earth, general distance measure
$C_{\Delta g}$	systematic error in a gravity observation
C	a constant corresponding to pressure for isostatic equilibrium
d	general distance measure
D	depth of compensation in computations of isostatic anomalies
$E(x, y)$	operator for the sum of the products of vectors x, and y
f	expression used to indicate a function

$\mathbf{F}$	vector of force
g	magnitude of the attraction of gravity
g	$(g_x, g_y, g_z) =$ vector direction of gravity
$\mathbf{\Delta g}$	$(\Delta g_x, \Delta g_y, \Delta g_z) =$ is the gravity anomaly, or gravity value
Δg_{Bp}	the Bouguer plate correction, anomaly from a flat plate
Δg_B	simple Bouguer anomaly
$\overline{\Delta g}$	mean value of gravity anomaly
G	universal gravitational constant, $6.67428 \pm 0.00067) \times 10^{-11}$ m^3 kg^{-1} s^{-2}
h	height, as above sea level, or linear measure in vertical direction
k	linear spring constant, Young's modulus
$K(r)$	is the autocovariance function of distance r
l	distance between two attracting masses, general length measure
L	measure of length
$m_{\Delta g}^2$	variance (error) of a gravity observation
m_x	variance of measurement of x
m	attracted mass
m'	attracting mass
M_m	mass of the Moon
M_e	mass of Earth
$M_{\Delta g}^2$	variance of gravity data, including reading errors and systematic errors
M	gravitational restoring moment
$\overline{M}$	internal restoring moment
N	the geoid, shape of the Earth
n	normal to a surface
P	pressure
r	radius or position vector, radial coordinate in spherical coordinates
r_e	radius of the Earth
R	radial distance from origin, radius of Earth
s	measure of distance, constant in smoothing operator
t	time
T	total duration, period of oscillation, temperature
U	a potential function, or theoretical potential of the Earth
v	volume, velocity
V	Potential of attraction of a mass, or of the non-rotating Earth
W	A potential, on the Earth the gravitational potential including rotation
x	direction axis in rectangular coordinates that goes through 0 longitude
y	direction axis orthogonal to x, in Earth on the equatorial plane and perpendicular to the x-axis
z	the third axis in rectangular coordinates, axis of rotation of the Earth
(x, y, z)	position, position of test mass in gravitational field

References

Adams, Donald C., and Randy G. Keller, (1996). Precambrian basement geology of the Permian basin region of west Texas and eastern New Mexico: a geophysical perspective, *Bulletin of the American Association of Petroleum Geologist*, Vol. **80**, 410–431.

ASTM Standard, D6430, 1999 (2010), "Standard guide for using the gravity method for subsurface investigation", ASTM International, West Conshohocken, PA, 2010, DOI: 10.1520/D6430-99R10.

Birch, F., (1961). The velocity of compressional waves in rocks to 10 kilobars. Part 2. *Journal of Geophysical Research*, Vol. **66**(7), 2199–2224

Blakely, Richard J., (1995). *Potential Theory in Gravity and Magnetic Applications*, New York, Cambridge University Press, New York, p. 441.

Brocher, T. M., (2005). Empirical relations between elastic wavespeeds and density in the Earth's crust, *Bulletin of the Seismological Society of America*, Vol. **95**(6), 2081–2092.

Butler, D. K., (1984). Microgravimetric and gravity gradient techniques for detection of subsurface cavities, doi: 10.1190/1.1441723, *Geophysics*, Vol. **49**(7), 1084–1096.

Cogbill, A. H., (1990). Gravity terrain corrections using digital elevation models. *Geophysics*, Vol. **55**, 102–106.

Cook, Frederick A., Larry D. Brown, Sidney Kaufman, Jack E. Oliver, and Todd A. Peterson, (1981). COCORP seismic profiling of the Appalachian orogen beneath the coastal Plain of Georgia, *Geological Society of America Bulletin*, Part 1, Vol. **92**, 738–748.

Dobrin, Milton B., (1976). *Introduction to Geophysical Prospecting*, New York, New York, McGraw-Hill Bock Company, New York, New York, p. 630.

Dorman, L. M., and B. T. R. Lewis, (1970). Experimental isostasy, 1. Theory of the determination of the Earth's isostatic response to a concentrated load, *Journal of Geophysical Research*, Vol. **75**, 3357–3385.

Dorman, L. M., and B. T. R. Lewis, (1972). Experimental isostasy, 3, Inversion of the isostatic Green function and lateral density changes, *Journal of Geophysical Research*, Vol. **77**, 3068–3077.

Fedi, Maurizio, and Mark Pilkington, (2012). Understanding imaging methods for potential field data, *Geophysics* Vol. **71**(1), G13–G24.

Gardner, G. H. F., Gardner, L. W., and Gregory, A. R., (1974). Formation velocity and density – the diagnostic basics for stratigraphic traps. *Geophysics*, Vol. **39**(6), 770–80.

Hammer, S., (1939). Terrain corrections for gravimeter stations: *Geophysics*, Vol. **4**, 184–194.

Hawman R. B., (2008). Crustal thickness variations across the Blue Ridge Mountains, Southern Appalachians: An alternative procedure for migrating wide-angle reflection data, *Bulletin of the Seismological Society of America* Vol. **98**(1), 469–475.

Hawman, Robert B., Mohamed O. Khalifa, and M. Scott Baker, (2011). Isostatic compensation for a portion of the Southern Appalachians: Evidence from a reconnaissance study using wide-angle, three-component seismic soundings. *Bulletin of the Seismological Society of America*, Vol. **124**(3–4), 291–317.

Heiskanen, W. A., and F. A. Vening Meinesz, (1958). *The Earth and its Gravity Field*, McGraw-Hill Book Company, p. 364.

Heiskanen, W. A., and H. Moritz, (1967). *Physical Geodesy*, San Francisco, W. H. Freeman and Company, San Francisco, p. 364.

Heiskanen, Weikko A., and Helmut Moritz, (1965). *Physical Geodesy*, San Francisco, W. H. Freeman and Company, San Francisco, p. 364.

Hinze, W. J., Aiken, C., Brozena, J., Coakley, B., Dater, D., Flanagan, G., Forsberg, R., Hildenbrand, T., Keller, G. R., Kellogg, J., Kucks, R., Li, X., Mainville, A., Morin, R., Pilkington, M., Plouff, D., Ravat, D., Roman, D., Urrutia-Fucugauchi, J., V´eronneau, M., Webring, M., and Winester, D., (2005). New standards for reducing gravity data: The North American gravity database, *Geophysics*, Vol. **70**(4), J25–J32.

Kaufmann, R. D., and L. T. Long, (1996). Velocity structure and seismicity of southeastern Tennessee, *Journal of Geophysical Research*, Vol. **101**(B4), 8531–8542.

LaFehr, T. R., (1991). An exact solution for the gravity curvature (Bullard B) correction: *Geophysics*, Vol. **56**, 1179–1184.

Lees, J. M., and J. C. VanDecar, (1991). Seismic tomography constrained by Bouguer gravity anomalies: Applications in Western Washington, *PAGEOPH*, Vol. **135**, 31–52.

Lewis, B. T. R., and L. M. Dorman, (1970). Experimental isostasy, 2, An isostatic model for the U.S.A. derived from gravity and topographic data, *Journal of Geophysical Research*, Vol. **75**, 3367–3386.

Long, L. T., and Jeih-San Liow, (1986). Crustal thickness, velocity structure, and the isostatic response function in the Southern Appalachians, in Reflection Seismology: *The Continental Crust, Geodynamics Series*, American Geophysical Union, Vol. **14**, 215–222.

MacMillan, W. D., (1958). *The Theory of the Potential*, New York, Dover, New York, p. 469.

Menke, William, (1984). *Geophysical Data Analysis: Discrete Inverse Theory*, New York, Academic Press, Inc., New York, p. 260.

Mohr, P. J., B. N. Taylor, and D. B. Newell, (2011). "The 2010 CODATA Recommended Values of the Fundamental Physical Constants" (Web Version 6.0). This database was developed by J. Baker, M. Douma, and S. Kotochigova. Available at http://physics.nist. gov/constants [Friday, 22-Jul-2011 10:04:27 EDT]. National Institute of Standards and Technology, Gaithersburg, MD 20899.

Morelli, C., ed., (1974). *The International Gravity Standardization Net, 1971*, International Association of Geodesy, Special Publication 4.

Moritz, H., (1980). Geodetic Reference System 1980, *Journal of Geodesy*, Vol. **54**, 395–405.

Nafe, J. E., and C. L. Drake, (1963). Physical properties of marine sediments. In *The Sea*, Vol. **3**, ed. M. N. Hill. New York, Interscience, New York, pp. 794–815.

Nagy, D., (1966). The gravitational attraction of a right rectangular prism, *Geophysics*, **31**, 362–371.

Natonal Geospatial-intelligence Agency, Office of GE OINT Sciences: (2012). Gravity Reference Base Station File History. http://earth-info.nga.mil/GandG/geocontrol/rbshist.html.

Nettleton, L. L., (1939). Determination of density for reduction of elevation factor. *Geophysics*, Vol. **4**(3), 176–83.

Parks, H. V., and Faller, J. E., (2010). A simple pendulum determination of the gravitational constant, *Physical Review Letters*, Vol. **105**, Issue 11, id. 110801. http://link.aps.org/doi/10.1103/PhysRevLett.105.110801.

Rothe, George H., and L. T. Long, (1975). Geophysical investigation of a diabase dike swarm in west-central Georgia, *Southeastern Geology*, **17**(2), 67–79.

Stephenson, Randell, and Kurt Lambeck, (1985). Isostatic response of the lithosphere with in-plane stress: application to central Australia, *Journal of Geophysical Research*, Vol. **90**(B10), 8581–8588.

Talwani, M., and M. Ewing, (1960). Rapid computation of gravitational attraction of three-dimension bodies of arbitrary shape. *Geophysics*, **25**(1), 203–225.

Telford, W. M., L. P. Geldart, and R. E. Sheriff, (1990). *Applied Geophysics*, 2nd Edition, New York, Cambridge University Press, New York, p. 770, table 6.1 Average densities of common sedimentary rocks.

Watts, A. B., (2001). *Isostasy and flexure of the lithosphere*, Cambridge, United Kingdom, Cambridge University Press, Cambridge, United Kingdom, 451 pp.

Index